Montana's Gallatin Canyon

A Gem in the Treasure State

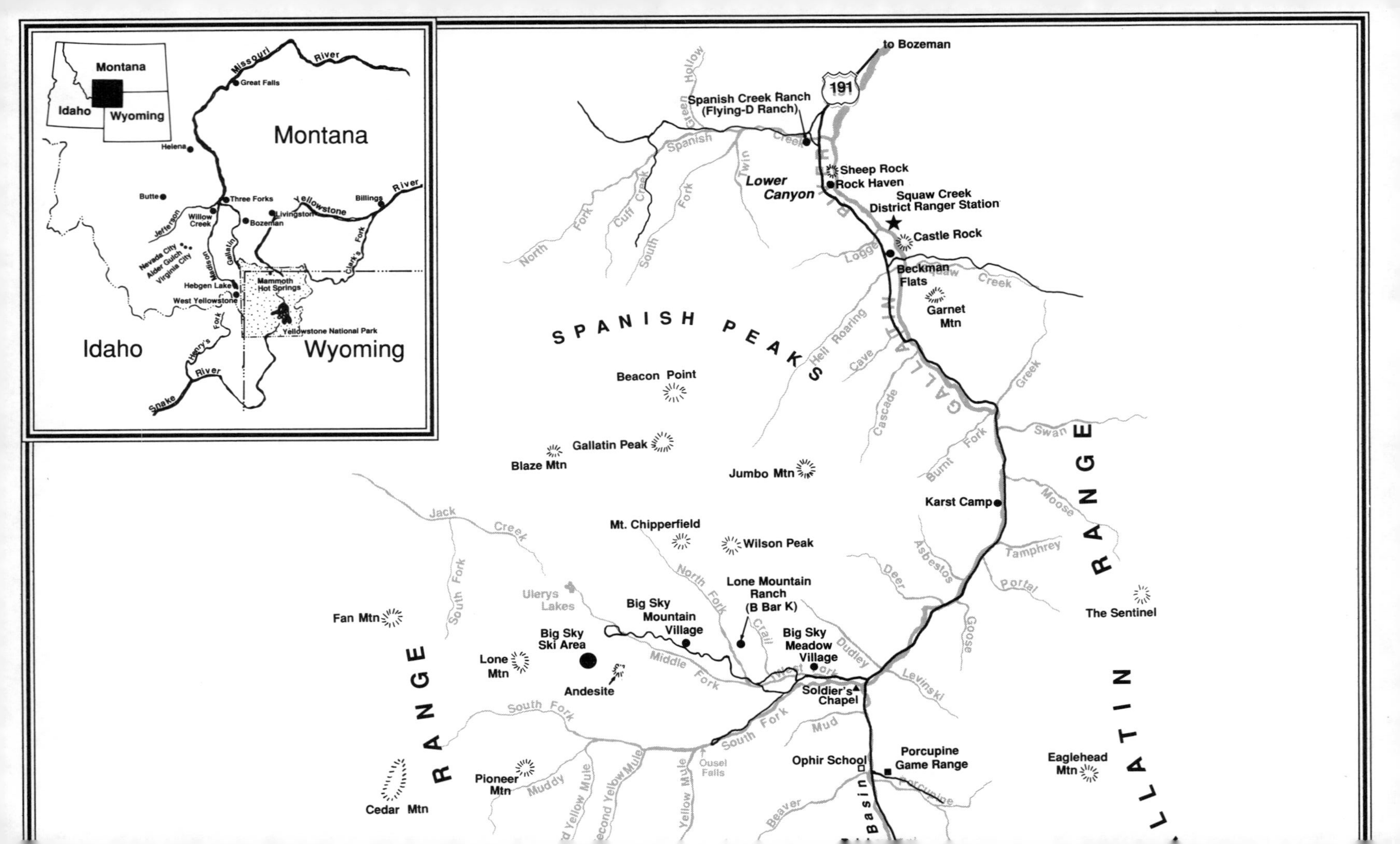

Montana
Idaho
Wyoming
Montana
Missouri River
Great Falls
Helena
Butte
Three Forks
Willow Creek
Bozeman
Livingston
Yellowstone River
Billings
Clark's Fork
Jefferson
Gallatin
Madison
Nevada City
Alder Gulch
Virginia City
Hebgen Lake
West Yellowstone
Mammoth Hot Springs
Yellowstone National Park
Idaho
Wyoming
Henry's Fork
Snake River
to Bozeman
191
Spanish Creek Ranch
(Flying-D Ranch)
Sheep Rock
Rock Haven
Lower
Canyon
Squaw Creek
District Ranger Station
Castle Rock
Beckman
Flats
Garnet
Mtn
SPANISH PEAKS
Beacon Point
Gallatin Peak
Blaze Mtn
Jumbo Mtn
Karst Camp
Mt. Chipperfield
Wilson Peak
Lone Mountain
Ranch
(B Bar K)
Big Sky
Mountain
Village
Big Sky
Meadow
Village
Big Sky
Ski Area
Lone
Mtn
Andesite
Soldier's
Chapel
Fan Mtn
Ulerys
Lakes
Pioneer
Mtn
Cedar Mtn
Ophir School
Porcupine
Game Range
Eaglehead
Mtn
The Sentinel
GALLATIN RANGE
RANGE
Green Hollow
Spanish Creek
Twin
Cuff Creek
North Fork
South Fork
Logger
Hell Roaring
Cave
Cascade
Greek
Swan
Burnt Fork
Moose
Tamphrey
Portal
Asbestos
Deer
Goose
Levinski
Dudley
Crail
North Fork
Middle Fork
West Fork
Jack Creek
South Fork
South Fork
Mud
Ousel
Falls
Yellow Mule
Second Yellow Mule
Muddy
Beaver
Porcupine
Basin

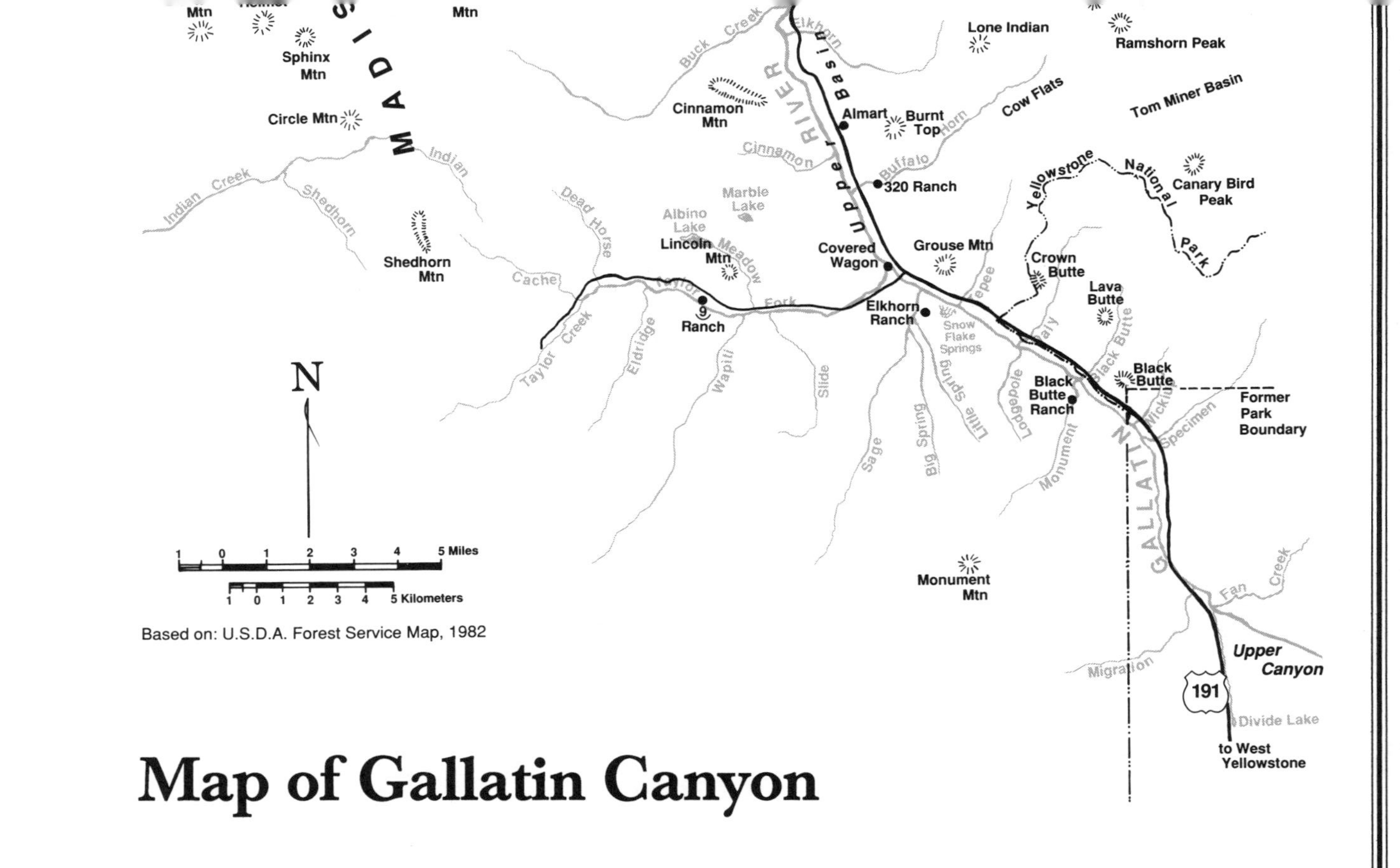

Based on: U.S.D.A. Forest Service Map, 1982

Map of Gallatin Canyon

Montana's Gallatin Canyon
A Gem in the Treasure State

Janet Cronin
Dorothy Vick

Mountain Press Publishing Company
Missoula, 1992

Cover Art by Allan Kass

Maps by Trudi Peek

Unless otherwise credited, all
photographs courtesy of the Michener family.

Library of Congress Cataloging-in-Publication Data

Cronin, Janet
Montana's Gallatin Canyon : a gem in the Treasure State / Janet Cronin, Dorothy Vick.
p. cm.
Includes bibliographical references and index.
ISBN 0-87842-277-3 : $12.00
1. Gallatin Canyon (Mont.)—History. 2. Gallatin Canyon (Mont.)—Social life and customs. I. Vick, Dorothy, 1914- . II. Title.
F737.G2C76 1992 92-36887
978.6'662—dc20 CIP

Mountain Press Publishing Company
P.O. Box 2399 • Missoula, MT 59806
406-728-1900 • 800-234-53008

For Raymond and Francis
and
for Paul

Table of Contents

Foreword

Future residents of the Gallatin Canyon and the steadily increasing number of visitors to the Yellowstone National Park area will be indebted to Dorothy Michener Vick and Janet Cronin for this first history of the canyon.

The near ninety-mile length of the Gallatin River and its major tributaries flow between towering peaks and open grassy parks now largely settled in dude ranches and the home ranches of the extensive livestock industry, such as the enormous Flying D Ranch, now the Ted Turner buffalo experiment, as well as Chet Huntley's Big Sky Ski resort.

The long-delayed modern highway that parallels the river is an important one, bringing Yellowstone National Park some of its two million visitors each year.

Mrs. Vick is unusually well qualified to write a first history, since she and her pioneering family have been a part of the development of the canyon for many years.

In straightforward, concise style, the authors have woven a fascinating story of life in the canyon. The legendary Pete Karst takes on full stature in the description of his extensive tourist camp and its burning. The tragic Andrew Levenski incident becomes more understandable with the background of the Hercules Mining Company. Informative and enjoyable, I highly recommend this book.

—Merrill G. Burlingame

Dr. Caroline McGill

Preface

In 1938 Dr. Caroline McGill, who had recently acquired the 320 Ranch, began collecting the history of the Gallatin Canyon. She wrote letters to many people who had lived in the area around the turn of the century, asking them to send her written recollections and photos of their early years. Many people responded to the doctor's inquiries. She also visited with early settlers and talked to them about the first homesteaders. She asked where they settled, how they had lived, and how places got their names. Late at night Dr. McGill recorded their stories on paper in her illegible hand; later her secretary transcribed them. A good interview with miner and dude rancher Pete Karst resulted from the doctor's efforts, as does Foley Water's written account of early homesteading.

The doctor involved Vic Benson of the Covered Wagon Ranch in her project. Vic spent two years traveling up and down the canyon, visiting with friends and asking questions. He preserved names, dates, and places. His wife, Eda, spent many hours typing the notes her husband made.

Dorothy Michener Vick returned to the canyon as the doctor was collecting her history. Vick's father, Thomas Michener, had started a history of the area in 1917 but died before finishing his work, and Dorothy's mother encouraged her to complete the project. When Dr. McGill urged Dorothy to assist in assembling information on the canyon, Dorothy began to save every newspaper reference to the area. She preserved official documents and other written information, taped interviews with early residents, and recorded her own memories.

The results of these efforts reside in the Special Collections Room of the Montana State University Library.

The information generated by Dr. Caroline McGill's enthusiasm and persistence made this book possible.

Acknowledgements

The authors would like to give special thanks to Francis Niven for sharing the results of his many years of historical research. His generosity saved us many tedious hours poring over old newspapers.

We would also like to thank the staff of the Merrill G. Burlingame Special Collections of Montana State University. Nathan Bender, Marjorie David, and Deborah Nash truly made us feel that their one goal was to make our project successful. We now understand why authors so often profusely thank research librarians.

Many other people contributed to our project. Dr. Merrill Burlingame read the manuscript and we thank him for reviewing our work. The people at the U. S. Forest Service, especially Hillary Dustin, provided us with information on the Gallatin National Forest, while Walt Allen and Thelma Michael gave us access to old photos. The staff and volunteers at the Gallatin Valley Pioneer Museum provided information and photos. Anna Price tirelessly searched out reference materials.

Other people helped us in many ways. We wish to thank Margaret Michener Kelly Thorsen, Ora and Stanley Lemon, Helen Michener Kane, Raymond T. Michener, Patricia Goodrich and her daughter Gail Goodrich Walma, Sara Durnam Anderson, Buck and Helen Knight, George and Clara Lemon, Susie Taylor, Elaine Hume, Don Corcoran, Chuck Anceney, Dorothy Nile, L. Lawrence, Bob Evans, Ralph Paugh, Joan Wilsey, Peggy Todd, Bill Penttila, Jean and Bob Simkins, Mary Owens, Tom Tankersley, Rick Harrison of Rikshots, Lela Wilkins, June Benham Stricklin, Jack Duke, Barbara Hymas, Jeanne Powell, Kevin Sanders, Mary Lou Cook, and Tod King.

View of old road near Castle Rock in lower canyon, 1927.

A Gem in the Treasure State

Cradled in the mountains of southwestern Montana, the Gallatin Canyon is characterized by a wild beauty. A small gem in the vast Treasure State, the canyon offers scenery from rugged mountains to rolling foothills to wide meadows. Yet the terrain that gives the canyon its rich natural beauty also made it virtually inaccessible until the beginning of the twentieth century.

The early settlers, who negotiated the rocky trail and later the narrow road to gain access to the canyon, had to wrest a living from an unfriendly topography—the small mountain meadows, glacier-carved cirques, and imposing mountains meant a dearth of flat land for farming. Once established in a log cabin the settler faced the deep snow and sub-zero temperatures typical to high-elevation winters. Today these conditions make the canyon a mecca for winter-sports enthusiasts and an ideal setting for big game hunters. But for the early settler, such weather added to the problems faced by neighbors in lower elevations.

The difficult access, lack of arable land, and extreme weather conditions kept the population small and limited exploitation of the canyon. Cattle grazing and logging began in the 1880s and continue today, yet the canyon remained largely isolated from the outside world until the late 1960s, when the development of the Big Sky Resort changed the tempo of the area.

Events that shaped the surrounding areas of the state had less effect on the canyon. We know fur trappers traveled through the area, because they noted the rivers and mountains on early maps. Yet no fur trading forts sprang up in the remote mountain valley. Prospectors undoubtedly

Rock spires near Cave Creek. —Dorothy Nile

combed every corner of the canyon, but they discovered no extensive deposits of gold, which spared the area a large influx of miners and the supporting businesses that follow. The building of the transcontinental railroad just north of the canyon created a demand for railroad ties. Many men traveled up the Gallatin River to fell logs for ties, but the men disappeared after completing the railroad. Farsighted businessmen from Bozeman surveyed the canyon hoping for a railroad link to the nation's first national park. But in the end tourists reached Yellowstone National Park by rail via the Yellowstone Valley and West Yellowstone, sparing the canyon the intrusion of a rail line. Homestead laws opened the area to early settlers, but only a hardy few survived the isolation and severe winters of the canyon.

Gallatin County finally constructed a road into the canyon in 1898, thirty-four years after settlers founded the town of Bozeman. Progress followed the road: the telephone early in the twentieth century, road improvements in 1927, 1939, and the early 1950s, and electricity in 1949. Even so, the canyon remained a quiet, neglected little corner of Montana, known to the few who kept its charms a secret until Chet Huntley and the Big Sky Resort brought fame and fortune to the area.

This book focuses on the early days, the days of the trappers and prospectors, the loggers and cattle men, the dude ranchers and hunters, and most of all, the homesteaders. Descendants of early settlers still live in the canyon, a tribute to the hard work of their forbearers and to their love of this special place. We hope this book brings to life the story of the early years of discovery, labor, hardship, success, disappointment, and fulfillment.

U.S. Highway 191 runs through the Gallatin Canyon along the Gallatin River and connects Bozeman, situated in the Gallatin Valley to the north, with the town of West Yellowstone to the south. The canyon stretches approximately fifty miles.

Entering the area from the north, you travel across the wide, fertile Gallatin Valley toward rolling slopes and abutting mountains. Slipping through an opening in the hills, you find yourself enclosed in the Gallatin Canyon. For perhaps ten miles the canyon floor is wide. The large Spanish Creek drainage leads off to the west. If you follow the dirt road along Spanish Creek to the Spanish Creek Campground, you gain access to the north side of the Spanish Peaks. If the road through this valley were open to the public, you could drive on through to the Madison River.

Continuing south on U.S. Highway 191, you pass Sheep Rock and then Castle Rock, tall limestone formations carved by wind and water into imposing edifices. After you pass Hell Roaring Creek the canyon narrows, with steep tree covered cliffs and rock slides falling at abrupt

Cinnamon Mountain from Buffalo Horn Creek. —Dorothy Nile

angles to the river. Here the Shoshone name for the Gallatin, *Cut-tuh-o-gwa*, Swift Water, accurately describes the river.

As you continue up the canyon the river runs between sheer walls and rock slides for the next fifteen miles. Before the days of modern road equipment, a narrow trail hugging the side of the mountain provided the only access to the upper canyon. Extensive road work has narrowed the stream bed, making way for the highway that we use today. Gone are Sage Brush Point, where the road looked down 150 feet to the river, and the pole bridges that spanned the rushing Gallatin.

The terrain opens at West Fork, forming the lower basin, which runs for about five miles. Past Twin Cabins Creek the mountains again crowd the river into a narrow channel, which opens at Cinnamon Creek to form the upper basin. This basin runs south into Yellowstone Park. At tiny Divide Lake, at the southern end of the upper basin and fifty miles from the canyon entrance, you leave the canyon and travel on to West Yellowstone.

Born a tiny stream in the northwest corner of Yellowstone National Park, the Gallatin River flows north through canyon and valley. It joins the Madison and Jefferson rivers to form the Missouri River. On its run through the canyon, the river cuts through sedimentary, igneous, and metamorphic rock. Where the river met unresisting rock, it carved the lower and upper basins. The tributaries of the Gallatin River in these areas of soft rock also carved wide valleys. These include the West Fork drainage and the Porcupine Creek area in the lower basin, and Tepee and Daly creeks in the upper basin. In the areas where the water met unyielding metamorphic rock, the canyon narrows: stoney slopes and rock slides constrict the river, which carves the canyon deeper with each passing year.

During periods of glaciation, grinding ice further widened and deepened the canyon and some of the drainages that flow into it. Water and ice have left the canyon varying in width from several hundred feet to three-quarters of a mile wide. The elevation decreases from 7,600 feet at the southern end to 5,000 feet at the northern end.

Higher and narrower than the parallel Madison and Yellowstone river valleys, the canyon is separated from these valleys by the Madison Range on the west and the Gallatin Range on the east. Creeks and streams of varying size course down from these two mountain ranges. Some crash precipitously down the steep, rocky walls of the canyon. Others wind gently through wide, open meadows before feeding the Gallatin River as it flows north.

Upper basin near Sage Creek. —Dorothy Nile

Deviants in this area of north-south ranges, the upthrust Spanish Peaks run east to west at the northern end of the canyon. These peaks are about the same age as the sedimentary Madison Range and the igneous Gallatin Range but are higher and more rugged because the gneiss that composes them is hard and resistant to erosion. The Spanish Peaks include 11,015-foot-high Gallatin Peak and 10,700-foot-high Wilson Peak. The honor of highest peak in the area goes to the 11,166-foot volcanic remnant known as Lone Mountain. Part of the Madison Range, it dominates the West Fork drainage, site of the Big Sky Resort.

Evergreens cover the mountains, while grasses and sagebrush carpet the rolling hills and meadows of the canyon. Willows and dogwood line the river and streams, providing cover for moose and other animals. In addition to an assortment of small fauna, elk, deer, mountain goats, bighorn sheep, black bears, and grizzlies live in the area. The Gallatin River is a blue-ribbon trout stream much favored by fly fishing enthusiasts.

Looking north from Crown Butte to the Spanish Peaks. —Dorothy Nile

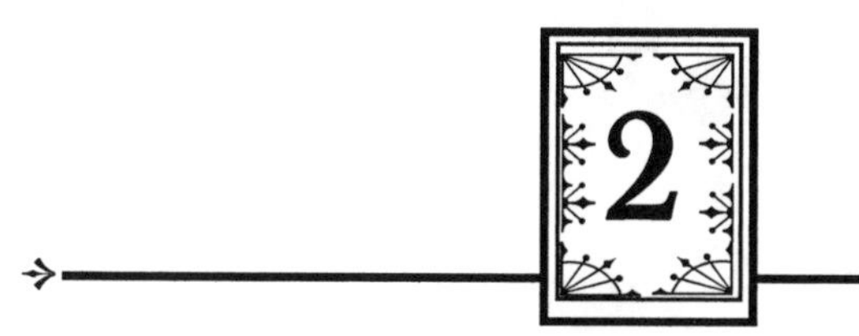

Native Inhabitants

From artifacts found in Montana, we know that early Indians lived to the northwest and northeast of the canyon as long ago as 10,000 B.C. Archaeologists have found artifacts dating from 9,000 B.C. in the Gallatin Canyon.

Our best information on prehistoric Indians in the canyon comes from a master's thesis, "Indian Occupation in the Gallatin Valley and Canyon," completed by Lewis Napton at the University of Montana in 1965. Napton excavated, first with his parents and then with university survey crews, for many years. He surveyed Indian occupation sites along the length of the canyon. He divides the Indian occupation into three periods that he distinguishes by projectile points, chipped stone points used on darts or spears. The occupation periods he identifies coincide with climatic changes.

During the early period, from 10,000 to 5,000 B.C. Indians used stemmed projectile points. This period began at the end of the last ice age in North America. As the ice retreated, it left the northern Rockies cold and wet with tree-covered mountains. Several projectile points found in the Gallatin Canyon date from this period.

Corner-notched projectile points distinguish the middle period, which lasted from 5,000 B.C. to A.D. 500. During this warmer and drier period, expansive grasslands and occasional forests covered the area. The great plains to the east of the Rocky Mountains experienced drought, which drove many of the plains buffalo into the mountains in search of food. The Indians who depended on these animals for food, clothing, and shelter followed in pursuit. Napton considers this the time when the greatest

Projectile points and scrapers collected in Gallatin Canyon.

number of Indians used the canyon, and most artifacts recovered have been from this period.

The late period covers the last two to three thousand years. Side-notched projectile points represent this period. A change to a climate similar to the one we have today began about three thousand years ago. The great plains to the east once again received enough moisture so the grass grew tall and lush, and the buffalo herds increased. The Indians, once forced by drought into the mountains, flowed back onto the plains where they stayed until the westward march of the white man ended forever their way of life. Napton found relatively few artifacts from this period in the canyon.

The small, seasonal campsites found by Napton suggest that in all three periods family groups and bands of hunters used the canyon in spring, summer, and fall as a hunting ground. In prehistoric times, as well as today, big game moved from the lower valleys into the mountains surrounding the canyon during the summer. Artifacts found in high mountain passes of the Gallatin and Madison ranges suggest that Indian hunters used these trails to track migrating elk and mountain buffalo as the animals left the open, low land of the Madison and Yellowstone valleys and retreated to the mountains.

Napton found many artifacts associated with hunting: arrow heads, spear points, knives, awls, and scrapers. The survey found few implements used for gathering and preparing plant food. The Indians may have collected such food in the canyon, but harvesting tools could have been less durable than those associated with hunting.

Indians must also have used the canyon as a route to the Yellowstone area. The park has large deposits of obsidian, a glass formed when lava cools immediately after eruption. The tribes used this material for making arrowheads and spear points. Projectile points of Yellowstone obsidian dating from the late period have been found as far east as Iowa. Native Americans most likely valued obsidian as a trading commodity, as well as for making weapons and cutting implements.

The many occupation sites found in Yellowstone Park show that Indians used the area extensively, especially in summer when travel was easy. Besides collecting obsidian the Indians may also have used the hot springs in the park for baths or medicinal purposes.

Napton found an average of one prehistoric Indian site every five miles along the Gallatin River. With few exceptions the Indians located their camps near streams or springs. They left the sites unfortified, making them impossible to defend. From this arrangement Napton infers that "The absence of fortified sites does not necessarily demonstrate that warfare did not occur, yet the impression gained from an appraisal of the archaeological sites found in the Gallatin area is that the inhabitants of these sites were not greatly concerned for their safety."

The survey found a large site at Squaw Creek, where Napton located a quarry of red jasper and brown chert along with abundant artifacts, cores, and flakes from these materials. In addition Napton found two blue beads used by early traders in the West.

At Greek Creek, a site destroyed by highway construction in 1953, Napton found a few flakes and one artifact. Moving up the canyon, the survey examined sites at Portal, Moose, West Fork, Porcupine, Beaver, Twin Cabins, and Tepee creeks. The survey found one exception to these river sites. On a high pass on Cedar Mountain that connects the Madison and Gallatin rivers, Napton discovered an early-period campsite with flakes of basalt and chert.

Napton used much conjecture in analyzing and drawing conclusions from a small number of artifacts. At a site located south of the West Fork, the researchers found one side-notched point, one side-ground point, and several corner-notched points. They also found a beveled knife and an awl-scraper. Napton theorized that this site:

> might have been continuously occupied by different groups of hunters who used various types of projectile points and tools. The site could have been continuously occupied during the summer months. One

might suppose that the Indian men hunted in the hills, while the women remained at the site—a procedure practiced by many hunting and gathering aggregations. The scrapers and the "awl" or perforator were used in preparing hides for clothing and shelter. The presence of faunal remains indicates that animals were butchered on or near the site. Perhaps this was a "kill" site (the animals might have been killed while grazing); however, since the lithic inventory consists of flakes and various types of implements, the site might also have been the scene of domestic occupation. The variety of projectile points indicates that the site was serially occupied; therefore, it might have been used at different times for different purposes.

Wickiup in upper basin. —Gallatin County Historical Society

Napton often excavated just ahead of highway crews who were improving the road. Road work destroyed the sites at Moose, Portal, Greek, and Tepee creeks. Mining operations and construction destroyed the West Fork site, and a flood destroyed one other site in the late 1950s.

The upper Gallatin has produced evidence of more recent Indian occupation. Wickiups, or tepees, have been found at several locations, both in and outside of Yellowstone National Park. One location has three collapsed wickiups made entirely of lodgepole pine. The largest lodge consists of more than one hundred pine poles, all hewn with non-metal axes. This site is a considerable distance from water. Another site has three wickiups of aspen poles. One hundred thirty aspens compose the largest lodge, while seventy-five poles make up each of the two smaller lodges. A metal axe felled some of these poles, indicating the Indians constructed these lodges after white men penetrated the region. In each location the large lodge had a fire pit in its center, and the Indians erected these lodges away from the main Gallatin River. The Indians could have constructed them as war lodges or hunting lodges. The wickiups are in disrepair and surrounding trees have fallen on them.

Most of the Indians using the area around the Gallatin Canyon depended on the buffalo for food, clothing, and shelter. These tribes included the Blackfeet to the north of the canyon, the Crows to the east, and the Shoshone and Bannocks to the west and south. In addition to these buffalo users, the peaceful Sheepeater Indians, relatives of the Shoshone, survived in the Yellowstone area on bighorn sheep.

As buffalo became scarcer, the Indians who depended on these animals had to travel great distances to the remaining buffalo-hunting grounds. These remaining herds were located to the northeast and east of the Gallatin Canyon on the lower Yellowstone and on Clarks Fork of the Yellowstone rivers. The Shoshone and Bannocks lived on the Snake River Plain on the west side of the Continental Divide. They traveled many miles by horse from their homes near the Camas Meadows in Idaho to reach the buffalo, and their route through the mountains became known as the Bannock Trail. The trail crossed north of West Yellowstone where it connected with north-south trails from the Madison Valley and the Gallatin Canyon. Hunting parties traveling the Bannock Trail probably used the Gallatin Canyon to make quick scouting trips to the north to check the whereabouts of the enemy Blackfeet Indians. By 1885 the buffalo had disappeared from the plains, and the white man had pushed the Indians onto reservations. The Bannock Trail was abandoned. You can still trace parts of the trail in Yellowstone National Park.

An old Indian trail at the north end of the canyon led from the Madison River near McAllister and entered the canyon at Spanish Creek. When travelers reached the canyon they could go north or south. Another trail came from the Madison Valley over the divide at Indian Creek, following a trail down to Taylor Fork. The trail then went up Buffalo Horn Creek and down Tom Miner Basin to the Yellowstone Valley. These trails were probably less traveled than the well-known Bannock Trail, although they may be older.

From the white explorers perspective the Shoshone and Bannocks maintained a relatively high standard of living by traveling many miles to the buffalo grounds. Their relatives the Sheepeaters appeared to the whites to have lived in the Indian version of the Dark Ages. These

Sheepeater Indians. —Yellowstone National Park Photo Archives

Indians roamed through the Yellowstone area, including the Gallatin Canyon, hunting bighorn sheep that provided them with food, skins for clothing, and horns for making bows.

Observers have left conflicting pictures of these people. Captain Benjamin Bonneville met three Sheepeaters in 1835 in the Wind River Range. He described them as miserable, poor hermits without horses and using only bows with stone tipped arrows to hunt the game on which they lived.

Osborne Russell kept extensive journals of his life as a trapper in the Rocky Mountains. These memoirs provide us with much of the information that we have about the area in the early 1800s. In 1835, in the Lamar Valley of what is now Yellowstone Park, he encountered about twenty Sheepeaters, whom he describes as:

> neatly clothed in dressed deer and sheepskins of the best quality and . . . perfectly happy. They were rather surprised at our approach and retreated to the heights . . . but we succeeded in getting them to camp with us. Their personal property consisted of one old butcher knife, nearly worn to the back, two shattered fuses which had long since become useless for want of ammunition, a small stone pot and about 30 dogs on which they carried their skins, clothing, provisions, etc., on their hunting excursions. They were well armed with bows and arrows pointed with obsidian.

Trading with the Sheepeaters yielded Russell "a large number of elk, deer and sheep skins . . . of the finest quality and three neatly dressed panther skins in return for awls, axes, kettles, tobacco, ammunition, etc."

P.W. Norris, superintendent of Yellowstone Park, described the Sheepeaters in the 1870s as lacking horses, firearms, and even the most rudimentary knives or axes. According to Norris, the Sheepeaters, even at this late date, used obsidian tools to cut posts for their wickiups.

The Sheepeaters moved continually as they followed the bighorn sheep. They lived in crudely constructed brush or pole wickiups, sometimes covering the brush with skins or dirt. Large dogs pulled their belongings on travois. They skillfully made bows from the horns of the mountain sheep and successfully traded these bows to many of the neighboring tribes.

In 1917 Thomas Michener began a history of the Gallatin Canyon, left unfinished at his untimely death in 1921. Michener first came into the canyon in 1889. His account gives us our only source of information on the earliest days in the canyon:

> Stories have been told many times by Indians to the early day trappers of a fierce battle that was fought perhaps several generations ago between the Sheep Eaters and bands of several other tribes. These

other tribes all made war on the Sheep Eaters and kept them back on the high mountains where they lived on mountain sheep. This battle lasted several weeks in the mountains of the Gallatin about Sheep Rock and finally ended on Three Rivers Peak [In YNP, close to Big Horn Pass] where the Sheep Eaters made their last stand. They made a large stone fort, which can still be seen but they were wiped out to the last man. Jim Bridger and other trappers of that day had heard of this battle as it was described by the Indians.

I saw it [the fort] in 1891. . . . but no one has any idea of the year in which this battle was fought as it was long before the day of the white man.

In 1872 the Hayden Geological Party were camped near where later the twin cabins were built [Twin Cabin Creek]. Jack Bean was guide for the party and one day he rode on ahead up the canyon. When he came to the snow slide between Buck and Cinnamon Creeks, he found that a battle had just been fought between two tribes of Indians. One tribe had built breastworks [fortifications] out of the timber carried down by the snow slide. In after years this caught fire or was set afire by some vandal and no trace of it is left except a few burned sticks.

Outside of the battle between the Astorians and the Blackfeet on the Yellowstone divide there have been no regular battles between the Indians and the whites on the upper Gallatin. There have been a few fights of small importance such as two white trappers had a fight with a small bunch of Blackfeet on Porcupine Creek in the early 1860s. One trapper and one Indian were killed. The other trapper made his escape.

Nels Murray and his brother Andy trapped on the Gallatin River. Nels was an old time Indian fighter and was like Jim Bridger in that he told tales of his doings that had to be taken with a grain of salt. And many were the wild tales that he told of the things he did. A good many of them were probable while many were very improbable.

Nels and his brother Andy were camped on what is now the Wilson Ranch [320 Ranch today]. They had about 60 head of horses with them. Nels claimed that a few Bannock Indians were always on the lookout trying to steal the horses. One day some Indians camped near the mouth of Cinnamon Creek and Nels and Andy crawled up close to them through the sage brush and killed them all and threw their bodies into the river. The Indians never bothered them again but a disease broke out among their horses and they lost most of them in a few days. Geologist Peale named the creek where the horses died Murray Creek in honor of Nels Murray, but it was afterwards named Ramshorn and later Buffalo Horn, the name it still bears.

The Gallatin figures in the story of Nez Perce Chief Joseph, who led his tribe on a tragic march toward Canada and freedom. Michener gives this account:

> In the year of '77, the year of the Nez Perce out break, it was supposed that Joseph was making for the headwaters of the Gallatin. A party of soldiers was sent out from Fort Ellis. They went up the Yellowstone to Tom Miner Creek where they crossed the divide and came down Buffalo Horn Creek. They then crossed to the head of Tepee Creek and went down that creek to the Gallatin. They then followed Greyling Creek south only to find that the Indians with Howard on their trail had passed a few days before.

With the passing of Chief Joseph and his tribe, Indians disappear from the stories of the Gallatin Canyon. Only an occasional arrowhead, poking through the soil after a rain storm, remains from these early users of the land.

Three Forks of the Missouri River: The Gallatin, Madison and Jefferson.
—Gallatin County Historical Society

Beaver Trappers

Long before the Louisiana Purchase in 1803, Thomas Jefferson had shown an interest in exploring the upper Missouri River. The Louisiana Purchase gave President Jefferson the excuse to send the Lewis and Clark Expedition to investigate the newly acquired territory. Yet, there is much evidence that even had the United States not succeeded in acquiring the vast lands west of the Mississippi River, Jefferson still would have sent out the expedition. Jefferson envisioned the United States developing its own beaver trade in the Rocky Mountains. He also hoped to transport the beaver pelts trapped by the British and French in far away northwest Canada. At the turn of the century traders sent these furs to the east coast of Canada by canoe interrupted by lengthy portages. If Lewis and Clark could find routes to the Mississippi River from the Canadian north and a river route to the Pacific Ocean, the United States would become the major carrier of furs to Europe and to the Orient. In asking Congress for money to fund the Lewis and Clark Expedition, Jefferson tempted the legislators with the prospect of "great supplies of fur and peltry."

The Lewis and Clark Expedition left St. Louis, Missouri in the spring of 1804. It ascended the Missouri River to a Mandan village located in present-day North Dakota and wintered there. The expedition left the Mandan village in April of 1805. By late July it reached the headwaters of the Missouri River in the Gallatin Valley. This historic point is only thirty miles northwest of the Gallatin Canyon. On Sunday, July 28th Captain Lewis recorded the first information about the Gallatin River:

> Both Cap't C. and myself corrisponded in opinon with rispect to the impropriety of calling either of these streams the Missouri and

> accordingly agreed to name them after the President of the United States and the Secretaries of the Treasury and state having previously named one river in honour of the Secretaries of War and Navy. In pursuance of this resolution we called the S.W. fork, that which we meant to ascend, Jefferson's River in honor of that illustrious personage Thomas Jefferson. the Middle fork we called Madison's River in honor of James Madison, and the S.E. Fork we called Gallitin's River in honor of Albert Gallitin [the Swiss born secretary of the treasury]. The two first are 90 yards wide and the last is 70 yards. all of them run with great velocity and th[r]ow out large bodies of water, Gallitin's River is reather more rapid than either of the others, is not quite as deep but from all appearances may be navigated to a considerable distance. Capt. C. who came down Madison's river yesterday and has also seen Jefferson's some distance thinks Madison's reather the most rapid, but it is not as much so by any means as Gallitin's. the beds of all these streams are formed of smooth pebble and gravel, and their waters perfectly transparent; in short they are three noble streams.

The expedition proceeded on west without exploring the Gallatin River or the area from which it sprang. On their return trip from the West Coast in 1806, Captain Clark descended the Jefferson River to the the point where the river joined the Madison and Gallatin rivers and led his men over the pass later named for John Bozeman to the Yellowstone River. Captain Lewis took a more northerly route, and the two parties met near the confluence of the Yellowstone and Missouri rivers.

Lewis and Clark failed to locate a water route to the West Coast or a convenient way to transport furs from northern Canada, but they discovered an abundance of beaver in the Rocky Mountains. Beaver men invaded the northwest even as Lewis and Clark descended the Missouri River on their return trip to St. Louis in 1806. These trappers relied on the geographic information compiled by the expedition.

On 12 August 1806 the expedition met two trappers from Illinois, Joseph Dickson and Forest Hancock, at the junction of the Missouri and Yellowstone rivers. John Colter, a member of the Lewis and Clark Expedition, obtained permission to leave the group and join the two men in their quest for beaver. Colter eventually concentrated on the Three Forks region where the Gallatin, Madison, and Jefferson rivers meet and the Yellowstone Valley. Lewis and Clark identified both of these sites as prime beaver habitat. Both areas are close to the Gallatin Canyon.

The fur trade promised to be so lucrative that on 22 August 1806, as the Lewis and Clark Expedition made its way down the Missouri River, a chief of the Cheyennes asked Captain Clark to "Send some traders to them, that their country was full of beaver and they would then be encouraged to kill beaver, but now they had no use for them as they could get nothing for their skins and did not know well, how to catch beaver. . . . I promised the Nation that I would inform their Great father the President of the U States, and he would have them Supplied with goods. . . . " Clark made good on his promise, and American, French and English trappers soon overran the Rocky Mountains.

Lewis and Clark commented on the technique of trapping beaver. Meriwether Lewis, in January 1806, explained how to use the castors, scent glands located in sac-like tissue under the tail of the beaver, to make a potion with which to bait traps:

> Last evening Drewyer visited his traps and caught a beaver and an otter; the beaver was large and fat we had therefore fared sumptuously today; this we consider a great prize for the materials for making the bate [bait], the castor or bark stone is taken as the base, this is gently pressed out of the bladderlike bag which contains it, into a phiol [vial] of 4 ounces with a wide mouth; if you have them you will put from four to six [castor] stone in a phiol of that capacity, to this you will add half a nutmeg, a douzen or 15 grains of cloves and thirty grains of cinimon finely pulverized, stir them well together and then add as much ardent sperits to the composition as will reduce it to the consistency [of] mustard prepared for the table; when thus prepared it resembles mustard precisely to all appearance. when you cannot procure a phiol a bottle made of horn or a tight earthen vessel will answer, in all cases it must be excluded from the air or it will soon loose it's virtue; it is fit for uce immediately it is prepared but becomes much stronger and better in about four or five days and will keep for months

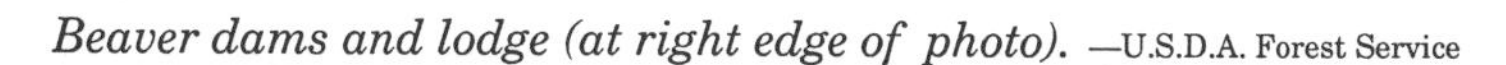

Beaver dams and lodge (at right edge of photo). —U.S.D.A. Forest Service

provided it be perfectly secluded from the air. when cloves are not to be had use double the quantity of Allspice, and when no spice can be obtained the bark of the root of sausafras. it appears to me that the principal uce of the spices is only to give variety to the scent of the bark stone and if so the mace vinellar and other sweet smelling spices might be employed with equal advantage.

Why was the fledgling government so interested in the fur trade with Europe and the Orient? The export of pelts would earn a considerable amount of foreign exchange for the United States. Since the seventeenth century, furs had been a source of income to the New World colonies. An estimate of the fur shipped from North America to Europe in the year 1770 puts the total value at $670,000.

By the early 1800s St. Louis had become headquarters for the companies which controlled the trappers to the west. Their operations alone were estimated to be worth $300,000 annually to the United States and trappers had barely begun to exploit the Rocky Mountains.

Although by 1800 the industrial revolution in Great Britain was well underway, the variety of luxury fabrics remained small and expensive. The demand for silk and other fine fabrics far outweighed the supply. People mainly wore homespun wool, cotton, or linen: stiff, itchy and drably colored. Fur provided a way to vary the wardrobe and signify social and economic status. Silky, light weight, and warm, fur made an elegant coat or stylish trim for collar, cuffs, or hem.

In addition to its use in its natural state, beaver fur lent itself to felting. The rough hairs of the beaver when steamed and pounded together create a type of cloth from which top hats were made. The bulk of the beaver pelts shipped from the United States and Canada were submitted to the felting process and ended up on a European gentleman's head.

What effect did the fur trade have on the Gallatin Canyon? We have many written accounts of trappers around the Three Forks of the Missouri River, on the Yellowstone River, on the Snake River Plains, and at Burnt Hole, now covered by Hebgen Lake. These areas surround the Gallatin Canyon, which trappers must have traversed as they made their way from one beaver ground to another. Yet the difficult access and the small amount of beaver habitat in the canyon compared to that in the surrounding areas kept trappers from using the area extensively.

German immigrant John Jacob Astor owned the American Fur Company, which made a legendary fortune in the Northwest. Thomas Michener, the early resident who started to write a history of the Gallatin Canyon, states that in 1812 trappers representing the American Fur Company were the first white men in the upper part of the canyon, possibly basing his account on written histories of the fur trappers:

> John Jacob Astor sent a group of men across country to the mouth of the Columbia River. They crossed the divide between Henry's Fork of the Snake and were ambushed on the Yellowstone divide between the Gallatin and Yellowstone rivers. Some of the party were killed but some escaped and were scattered. Joe Meeks, who afterward figured in Oregon history, got separated from all the rest and according to his story got back on the Gallatin and went up the stream to near its head and crossed the divide and came to a basin filled with vents of steam and spouting water, which he described as looking like the city of Pittsburgh, which he had left a couple of years before. This must have been Norris Geyser Basin.

Astor's men apparently followed the Bannock Trail from the south end of the canyon on through to the Yellowstone River.

The Crow Indians may have named the Spanish Peaks for a group of six or seven Spanish trappers whom they met in the area in 1838. U.S. Army Lt. James Bradley writes of information he learned from Little Face, one of his Crow scouts. Bradley questioned Little Face about geographical names in the summer of 1876, just before the battle of the Little Big Horn. The Indians called the mountains *"O O Ku shu-Ah Naht Si-Ah Rah Sah Ti,"* the canyon where the Spanish stop.

In his published journal, Osborne Russell, a trapper from Maine, mentions crossing the headwaters of the Gallatin in 1838. Tom Michener mentions a map drawn by the mountain man Jim Bridger in the 1830s. The map showed the upper Gallatin from its headwaters to the Porcupine and Beaver drainages.

Such is the fickle finger of fashion that those who had sported beaver hats at the start of the nineteenth century started wearing silk top hats in the 1830s, a happy coincidence for the few beaver remaining in the American West. The demand for beaver abated as their numbers declined. The mountain men no longer pursued the beaver with the intensity they had at the start of the century.

Gradually the trappers dwindled in number as demand shifted to buffalo hides for rugs and robes. A few hardy men who loved the mountains and the life they lived there continued to frequent the canyon with beaver traps. Now they also came with mule, pick, shovel, and gold pan. They combined trapping and prospecting and left their names and stories.

Eastern interests took the bulk of the money from the beaver trade out of Montana, a pattern followed over the years whether the commodity was furs, minerals, timber or railroad tariffs.

Cascade Creek. —Dorothy Nile

The Quest for Gold

In the winter of 1848 James Marshall found gold at Sutter's Mill, California, instantly changing the future of the western United States. Like blood hounds after a scent, prospectors spread out from the California gold fields. Gold strikes in Nevada, Idaho, and Colorado spurred miners to explore every inch of the Rocky Mountain region. In 1862 prospectors found a rich deposit of gold in the Beaverhead valley of Montana. Within a short time they established the gold mining town of Bannack. The race was on to find the next big lode.

Before the invention of mechanized mining equipment, prospectors used a more simple form of placer mining. Placer, a glacial or alluvial deposit of sand and gravel, contains eroded particles of valuable minerals. When snow, ice, rain, and wind erode gold-bearing rock, the detritus washes down into gulches and creeks. Gold weighs much more than rock and other minerals and settles at the bottom of a stream or gravel deposit. The gold eventually comes to rest on bedrock, although in gravel deposits the gold may lie several feet above the bedrock. If the stream carries the gold far from the mother lode, the force of the water tumbles the gold around, grinding it more and more finely. Eventually it becomes so light and fine that miners call it flour gold.

Placer mining requires little specialized equipment—just a pick, a shovel, and a gold pan, although large operations use more efficient and less labor-intensive methods. The prospector gets a pan of sand and gravel from just above bedrock, combines it with stream water and moves the pan in a circular motion. The swirling water carries the lighter gravel and sand over the edge of the pan while the gold settles to the bottom. When the lighter gravel washes away, black sand often remains in the bottom of the pan with the gold. Black sand is iron, a sister metal to gold, and they are frequently found together. Finding black sand excited prospectors because it meant gold, too, might be present. All gold mining in the Gallatin Canyon has been placer mining.

Most prospectors who worked placer deposits had little knowledge or experience at mining. Western lore has it that many prospectors used frying pans as they searched for the precious metal. If they did, they must not have cooked in them, or they found precious little: grease floats gold out of the pan while the gravel is being washed. Tom Michener, who prospected in the canyon for many years, warned people never to put their hands in his gold pan. He periodically burned his pan in a campfire to remove any grease residue.

Some prospectors spent years looking but never found gold. Often these men would get someone to grubstake them, paying for their food and other needs in return for half of whatever gold they found. Prospectors who could spin tales of the great riches awaiting discovery, who could describe what the next shovel full of gravel would contain, or who could imbue the life of a miner with the aura of romance, often succeeded in getting someone to back them. But all too often they lived on dreams.

Thomas Michener tells of an expedition in the spring of 1863. The group included Bill Fairweather, Edgar Clover, Barney Hughes, Harry

Black Butte in the upper basin. —U.S.D.A. Forest Service

Rogers, and two other men. Fairweather led the group of prospectors northeast from Bannack to the present site of Bozeman, then over the Bozeman Pass to the Yellowstone River. They planned to push on to the Big Horn Mountains where they intended to prospect for gold. They spent the night on the Yellowstone,

> where they were taken prisoner by the Indians and were robbed of all but their rifles. They were given some Indian ponies to take the place of their own good saddle horses and told to retrace their steps or they would be killed. The men crossed the Gallatin Range near the head waters of the Gallatin and camped at the foot of Black Butte on the Gallatin River. . . . Fairweather started to pan in the river and was just about through when some one cried that the Indians were coming. A group of young bucks had followed them in hopes of getting their scalps after all. Fairweather dumped his pan and ran for his horse.
>
> The group traveled fast until they were out of danger and two days later camped on Alder Creek [to the west of the Madison Range close by Virginia and Nevada cities]. Fairweather picked up a scad of gold [scads are small, flat pieces of gold] and one of the group panned out $2.40 worth of gold in his first pan. These ragged, half starved prospectors hurried to Bannack for supplies and were followed back to Alder Gulch by two hundred stampeders. Men came from every part of the earth to dig gold in Alder Gulch, and before the end of 1864 enough gold had been taken out with the crude methods of mining to pay for the Louisiana Purchase several times over.

More than six thousand people lived in the area around Alder Gulch by the end of the year. Many more combed every stream bed in the surrounding areas in the hopes of repeating the strike at Alder Gulch. The Gallatin Canyon is less than forty miles from the Alder Gulch area and prospectors unsurprisingly came into the canyon in search of their fortune.

Walter de Lacy prospected for gold in Alder Gulch in the summer of 1863. Born in Virginia in 1819, Walter Washington de Lacy received an extensive education. He spoke several languages and completed courses in mathematics, civil engineering and topography at West Point. De Lacy had a varied career, serving in the Mexican War and later in skirmishes against the Indians in the territory of Washington. In his capacity of civil engineer, de Lacy worked for railroads and territorial governments planning and surveying rail lines.

De Lacy's natural curiosity about the geography of the area in which he was prospecting and his desire to explore the upper Snake River to its source sent him on a six-week expedition. The journey ended with a trip down the Gallatin River. De Lacy and his group panned for gold along the streams they explored. De Lacy's diary holds no mention of panning for gold in the Gallatin River, but it does provide us a good description of the difficulty in traveling down the canyon:

> September 17th. To-day we passed through a cañon of granite, crossing the river several times, and having a rough time generally. In the evening we camped on a small flat, having made about fifteen miles, with a northeasterly course generally.
>
> On the 18th, we crossed to the left bank, where we discovered a trail, and in about a mile entered another granite cañon, which was the worst one that we had yet passed. The trail crossed the river many times, the fords were deep, and we had to climb over points of rock at every turn.

De Lacy's group passed Spanish Creek that same day, and Walter rode on to look out over the Gallatin Valley. He and his friends then went up Spanish Creek, prospecting as they went. Since they found no gold, they made their way back to Virginia City from which they had left more than a month earlier.

Lone Mountain and the Spanish Peaks. The Jack Creek road enters the Gallatin drainage of Lone Mountain —U.S.D.A. Forest Service

After his trip up the Snake River, de Lacy used his knowledge to create a map of the territory of Montana. He consulted with the territorial government and established the central starting point for the territorial survey at Willow Creek, near Three Forks.

Soon after the Alder Gulch strike, Bill Fairweather, one of the original discoverers, started the largest gold rush in the Gallatin Canyon. For reasons that remain unexplained, during the winter of 1863-64 Fairweather told a group of miners in Virginia City of panning for gold on Black Butte Creek in the Gallatin Canyon. He recounted getting a rich pan before Indians surprised him. Even though the weather was cold and the snow deep, two hundred men left Alder Gulch and started up Jack Creek for the Gallatin Canyon. Michener continues the story:

> A great snow storm caught them as they crossed the Madison divide, but instead of going back they pushed forward down the West Fork of the Gallatin hoping to reach the gold that Fairweather had told about. Nearly all of their horses starved to death. Think what a lust for gold it was that kept them working through the dead of winter over the pass between Jack Creek and the West Fork of the Gallatin.
>
> On the way to the Upper Basin they were overtaken by more prospectors who had come up the canyon from Bozeman. . . . The group of men camped at the foot of Black Butte on a small bar near the place where Fairweather had taken his pan. Very little prospecting was done as the ground was flat and full of water. The men had no idea of how deep they would have to go to hit bedrock. They held council and decided to go back to Alder. They had only found color and decided that Bill Fairweather was a damned liar. Their provisions were gone so they crossed the Gallatin Mountains and went down the Yellowstone Valley.

Other prospectors tried their luck in the canyon but in groups significantly smaller than the number who pursued Fairweather's promise in the middle of winter.

The story of the lost sluice box illustrates the continuing mania for gold that brought prospectors into the canyon. Michener again tells the story. An old man and his son came from Nebraska in 1864 to prospect in the Gallatin Range. In January or early February of 1865 they traveled down the canyon on homemade snowshoes. When they reached the Scarborough Ranch, a stage station on the road between Bozeman and Virginia City, the old man collapsed, unable to go on. The two men stayed with the Scarboroughs until the old man died. Coaches stopped daily at the ranch, so the story the two men told quickly reached the towns of Bozeman and Virginia City.

The young man, who was "not much more than half witted," claimed that they had left behind a crude cabin, sluice boxes and two horses that had starved to death. The father and son had taken about $1,000 worth of gold out of their diggings in the upper Gallatin, and the young man had the gold to prove it. After the father died the son packed up the gold and a sack of marten skins he had trapped near their abandoned cabin and took the stage to Nebraska.

The next summer the son returned with some friends to work the placer that he and his father had found. They came by steamboat to Fort Benton, just below the Great Falls of the Missouri. They then rode with a pack train into the Gallatin Canyon. The young man became confused, unable to find his way to the place where he and his father had found the gold. This started a hunt for the cabin and the lost sluice boxes that lasted for years.

Some of the Nebraska man's friends tried again to find the placer mine. They secured help from Virginia City prospectors, including John Donaldson. Donaldson told of the trip into the canyon and said that they found the old man's horses dead on the south slope of the mountains on the north side of the West Fork. Michener continues his story:

> Members of the group tried several times to find this lost Eldorado, but were never successful.
>
> John Reed and Pete Hartwick prospected on the West Fork in the place where the old man's boxes were supposed to have been, in '69-'72, but without success. They were good placer miners, having mined in Colorado, and they were in the first rush to Alder Gulch where they got good claims and worked them 'til the death of Hartwick who died sometime during the early '80s. . . . Old John Reed told me many times of the experiences he and his partner had. He always believed that somewhere on the West Fork was a large deposit of gold bearing gravel. He believed that when it was found, the source of the old lost sluice box story would be found. He claimed that he and Hartwick camped at the mouth of the West Fork in 1869 and dug a hole to bedrock above the rim which can be seen now. He says that they got twenty-five cents in their first pan. They got good pans afterwards but none so good as the first. They kept on up the West Fork prospecting but never found any more gold.

Thomas Michener also tells of his quest for gold and the lost sluice boxes:

> I first came to the Gallatin in 1889 with Jack Griffin. Happy Jack, as Griff was called, and I had two pack horses and our two saddle horses. At the mouth of the canyon we met Old Bob Dixon on his way back from Sun River. As Old Bob was acquainted with the upper country above the canyon better than any other man at that time and as he was one of the old time prospectors and trappers, we were glad to have his company. He had trapped and hunted in the company of Kit

> Carson . . . been in many Indian fights . . . and he was in the first rush to Alder Gulch.

Both Happy Jack and Old Bob told entertaining and instructive stories of the Old West. Michener goes on:

> The days I spent in their company in the years '89 and '90 are days that I have never regretted. Old Bob's trip to the Gallatin was mostly to trap bear but as Jack and I were after the lost sluice boxes he made that partly his business, too.
>
> Our first camp after coming through the canyon was on the West Fork about 100 yards below the mouth of the North Fork. We did very little digging but panned along every branch of the West Fork. This is the stream that we expected to find the lost sluice boxes on but they never materialized and we moved on to other streams.

Ousel Falls on the South Fork.
–U.S.D.A. Forest Service

> About this time and for the next five or six years there came into this country a miscellaneous lot of men composed of trappers, hunters, prospectors, cow men, and squatters. The streams were all thoroughly gone over. A preliminary prospecting was given to the whole country. I kept up the search for the lost sluice boxes. I tramped through every gulch, canyon, and stream between West Fork and Jackson Hole, Wyoming. I prospected all the mountains and studied geology. I did all that I could to get a knowledge of the country. I found lots of other minerals but the diggings of the old man and his son were still one of the lost mines.

Michener gives this account of the first cabin in the basin:

> In '75 two men came through from Oregon with pack and saddle horses. They made up their minds to put in the winter trapping on the West Fork. They built a cabin on the north side of the West Fork, about a mile above the mouth. The old cabin still stands [as of 1917] but is only a shell and most of the logs have rotted away. This was the first cabin ever built in the upper Gallatin. One of these men was named Lancing, but I never knew his partner's name. They did not stay all winter as the partner took a bad case of scurvy and had to be taken out.

Unlike many of the other prospectors who scoured the streams of the canyon and left, Michener stayed in the Gallatin. He proved up on a homestead, raised a family, ran a dude ranch, kept cattle and horses, continued mining operations, and tirelessly importuned the county government for better roads and a school.

The rush for gold pushed the people in the West to form territories, counties, and towns. Prospectors and settlers quickly recognized the need for local government to provide an orderly procedure for filing mining claims, enforcing property rights, maintaining law and order, and establishing water rights. Since placer mining requires large amounts of water, miners quickly filed riparian claims. To accommodate these needs the federal government formed the territory of Montana in 1864 and incorporated Gallatin County that same year.

The few miners tramping through the Gallatin canyon with gold pans and picks had little effect on the land. Pioneers in the surrounding areas, busy mining gold, setting up businesses, planting grain, and raising cattle, had little time to bother with the Gallatin Canyon. Even so, the settlements that sprang up in the southwestern part of Montana created a ripple effect that eventually reached the isolated canyon. The need for food in the Alder Gulch mining camps resulted in the development of the Gallatin Valley into farm land. With Bozeman as the county seat, the valley became prosperous on cattle and grain. The town of Butte—founded by gold miners, developed by silver, and brought to great wealth by copper—also became a market for the Gallatin Valley. Demand for the cattle and horses raised in the valley became so strong that by 1880 valley residents began to look around for more grazing lands. The Gallatin Canyon offered good summer pasture for cattle and horses. Despite the difficult access, valley farmers and ranchers soon began to pasture their herds in the Spanish Creek area and later in the basin. The canyon was about to change.

Cow moose. —Kevin Sanders

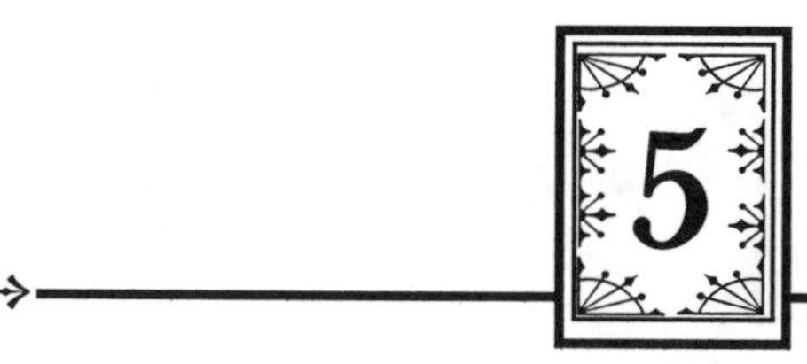

Early Hunters and Animal Tamers

Long after the early mountain men and beaver trappers traveled through the canyon, another group of outdoorsmen came into the area. These hunters and trappers of the last half of the nineteenth century traded the solitude, adventures, and hardships of their predecessors for an easier life. Most of the Indians had been pushed out of the area, towns had sprung up in the surrounding valleys, and supplies and companionship were relatively close. The canyon would never again be the untouched land that the first beavermen found.

Two of the earliest trappers were Nels and Andy Murray. In the 1860s these brothers hunted, trapped, grazed horses in the canyon, and terrorized any Indian they came across. The brothers came to Montana in a wagon train. On the journey Indians attacked the train and killed their older brother. For that reason Nels killed any Indian who crossed his path. According to one story, in the 1880s the brothers and some friends were staying with Jack Smith below the West Fork. They wanted to go to town to get supplies but were afraid to leave their belongings as three Indians were camped nearby. Nels quietly walked off and the men heard three shots. Nels then returned to say that it was safe to leave their belongings and go to Bozeman.

Another group of sportsmen used the canyon to combine the lifestyle they enjoyed with the opportunity to make a living. "Buckskin Charley" Marble and his wife Lizzie, Ira Dodge, and Dick Rock all hunted and trapped in the canyon. They also tamed and sold wild game. In 1892 the four hunters advertised in the Bozeman and

Hunters with Nels Murray (on right). —Museum of the Rockies Photo Archives

Gallatin Valley Directory as "Mountaineers, Guides and Collectors of Wild Animals."

We know little about Charles Marble's early days, except that he was a crack shot and dressed in buckskin clothes. Charley and Lizzie had a cabin up Taylor Fork on Meadow Lake, sometimes called Albino Lake, where they could stay year-round. To winterize their cabin they built double walls and insulated the space between them with dirt. A small lake, higher and to the west of their cabin site, is named Marble Lake. By the turn of the century Charley ran a taxidermy shop in Bozeman. He died in San Francisco in 1942 where he worked as a security guard in a war plant.

Marble kept a diary of his days in and around Yellowstone National Park. From it we know that in the fall of 1886 Buckskin Charley met Teddy Roosevelt in Bozeman and guided him on a forty-day hunting trip through the Gallatin Canyon. After buying the necessary supplies, they started by pack train over the difficult trail along the Gallatin River:

> As I remember it he was about thirty years old and wore glasses. I joked him about his glasses and told him he would have to take them off to kill his game. He said, "Show me the game."
>
> He had three guns but his "knockout," as he called it, was a Winchester 50, Explosive Express, '86 model, 90 grains of black powder, 505 grains of lead; in the end of the bullet was a round hole. In this was a .22 shell filled with black powder and countersunk flush with the end of the bullet. This would explode on hitting a sheet of cardboard or heavy writing paper. In elk and big game it tore a hole

as large as a hen's egg and seemed to have lots of penetration as well as shocking power.

We made camp on the Gallatin River just below the [Yellowstone] park line. While there we stacked all guns in camp and rode saddle horses into the park for several miles and took some pictures of mountain sheep, and elk, got one of a six-point close up in the act of bugling.

From there we went west up Sage Creek on the divide of the Gallatin and Madison Rivers. I planned on camping at Summit Lake, but finding a bunch of Crow Indians there I went on to the Devil's Bowl, at the head of Taylor's Fork. . . . We camped in the Devil's Bowl . . . by a little lake just at timberline . . . a couple of miles south of Shedhorn Ridge. This is the best mountain sheep country in the Gallatin and Madison Ranges. That evening at sundown a band of fourteen head came out on a rimrock a thousand feet above camp. An old ram hopped upon a pinnacle in the skyline and Teddy put the glasses on him and admired his big horns, and hoped to get him later. . . . The next day we went after sheep.

Buckskin Charley Marble.
—Francis Niven

The group climbed until reaching a point where the men could see the Teton peaks one hundred miles to the south, the Spanish Peaks to the north, and Electric Peak in Yellowstone to the east, and could even look down on the Continental Divide to the west. On going down the mountain they found a lion's den in a cave under a cliff littered with bones and the whole skeleton of a sheep the lions had killed and dragged in for their cubs.

> The next day I saw three rams coming over a ridge. . . . We picked our way carefully among the rocks and kept out of sight and above them as one had just as well stay in camp and read an almanac as to hunt uphill on mountain sheep for they are always looking down for the enemy, and by the time you get up to where they are they will be on the next high point looking back at you, and by the time you get there they are on the next one.

That day Teddy shot two trophy sheep.

> Next day we moved camp about eight miles north to the head of Indian Creek at the north end of Shed Horn Ridge. This is the camp where Teddy got his first bear. We had been hunting elk all day. On the way to camp about sundown we were riding through heavy timber. I showed Teddy fresh bear signs. We tied our horses and I told him there was a small lake near there and bears liked frogs as well as a bath once in a while.
>
> As we came in sight of the lake we noticed a commotion in the water, something swimming . . . towards us. Teddy said, "Is that a beaver?" "No, his head is too high, the ears too large, too wide across the head," I answered. "It's a bear. Get ready. Don't shoot until I tell you."
>
> He was a big black bear with that white spot under his throat. Soon he stopped and shook himself like a dog. I told Teddy to pull for the white spot. When that old .50 Winchester went off he rared up and fell over backwards, kicked and scrambled and bawled so one could hear him for a mile. But it was soon over. . . . We dragged him to the shore and managed to get him on the grass. Teddy then shook hands with me and said, "This is my first bear."

Teddy next shot an elk. Before the kill Marble had bugled and the elk had answered. This was something new to Teddy.

> He wanted to know how he made all that noise when he whistled or bugled. I showed him. An elk has no upper teeth in front, neither has a milk cow, but elk have two tusks [ivories] instead, as I showed him before we took them out. They use them when they whistle or bugle. They could not whistle or bugle without them. They curl up or pucker up their lips around the tusks to bugle. I have had in captivity tame

buffalo, moose, mountain sheep, elk and antelope as well as deer. It was then that I observed this characteristic about the elk. . . .

The Order of Elks used the tusks as an emblem for watch charms, stick pins and cuff buttons. This caused demand for them. This started about 1889. A good pair would sell for five dollars. About 1915 they would sell for fifty to seventy-five dollars. They were the next thing to legal tender, and hundreds of bull elk were killed for their tusks. Cows have them but not so large. This wanton slaughter went on for years, until at last Teddy Roosevelt discouraged the use of them as an emblem and now there is no demand for them which is a God's blessing for the elk and the young generation. . . .

Here the hunting party detoured to Targhee Pass and Cliff Lake for a few days of antelope hunting.

When we hit the old stage route at Ennis, twenty miles east of Virginia City, we took up Cedar Creek in the Madison Range, crossed the divide into Jack Creek, and camped several days beside the Spanish Peaks. . . . From this camp we crossed the divide and went down Spanish Creek. Here the beaver kept us awake again all night felling trees and splashing in the water.

Fifteen years later, when Roosevelt became president of the United States, the federal government still owned vast unsurveyed tracts of wilderness in the western states. Homesteaders eager to become property owners and entrepreneurs anxious to exploit the area's natural resources were settling the West. Roosevelt realized the value of large acres of wilderness: trees to harvest; watershed to benefit agriculture; grazing lands for cattle, horses, and sheep; wildlife for hunting and viewing; and that intangible benefit, recreation.

Teddy hunted big game, but he also enjoyed riding, camping, and viewing spectacular scenery. As president, Roosevelt had the opportunity to preserve federal lands for the benefit and enjoyment of all people. Did Teddy remember his hunting trip with "Buckskin Charley" when he signed the Gallatin National Forest into being? Regardless of whether he recalled his month in the Gallatin his years of enjoying the great outdoors must have influenced his decision to permanently preserve lands as national forests.

One of the more enterprising users of the canyon was Dick Rock, often called "Rocky Mountain Dick." Rock had guided for the U.S. Army and knew many early trappers and scouts, including Joe Meek, Kit Carson, Osborne Russell, and Jim Bridger. Rock shot with 45 Colt Shooters. He could toss two cocked guns into the air, then catch and fire them before they hit the ground.

Those who knew him describe Rock as tall, good looking, personable, and a loner who preferred to be in the mountains. Like many

—Fay Selby photograph

—Mrs. O.J. Salisbury photograph

Dick Rock and the buffalo that gored him to death.
—Photos from *Rocky Mountain Dick*, by Nolie Mumey (Denver: The Range Press, 1953); used by permission of Mrs. Norma L. Mumey.

mountain men, he could cover sixteen miles an hour on skis. He had six sled dogs that pulled his supplies over the snow on a toboggan. When President Grant first established Yellowstone Park, the administration allowed hunting, and Rock guided shooting parties into Yellowstone.

Rock had a cabin and corrals at Black Butte where he tamed elk and buffalo. He had a team of young bull elk that would pull a sled while in harness. He supplied elk, buffalo, geese, swans, mountain goats, and deer to zoos around the country. Shortly before his death in 1902 he sold a pair of mountain goats for $1,200. Rock had a bull buffalo that he rode. One day Rock entered the pen of the buffalo without the usual pitchfork in his hand, and the buffalo gored him to death.

Ira Dodge took out hunting parties in the upper Gallatin country. In 1893 Dodge and the George Marshall family took an exhibit of game animals to the Chicago World's Fair. Two corrals housed elk and deer, and Dodge and the Marshalls lived in a cabin constructed of logs that purportedly came from Yellowstone Park. Souvenir hunters at the fair whittled away at the cabin until nothing remained of it. Then one day a mad dog entered one of the corrals and bit an elk, which had to be put down. Despite these incidents the tour was a success. The Marshalls' son Frank served as game warden in the Gallatin Canyon in the early 1900s.

Early hunters and trappers like Dodge resented the laws that prohibited hunting and trapping in the park. In 1883 a construction company hired Dodge to provide meat for its crew, which was building a hotel in Yellowstone. Park authorities arrested Dodge for killing elk in the park. Later in his career a fight with a grizzly bear at Green River, Wyoming, left Dodge horribly mangled. He lived out his days at Pocatello, Idaho, minus most of his features.

Many of the names of the early hunters and trappers have been lost. Two trappers, their names forgotten, had a cabin on Porcupine in the late 1870s. Their crude, small log cabin, one of the first in the Gallatin Canyon, lacked windows. It had two small, flat whiskey flasks set in the door, either to let in light or to allow the inhabitants to see out. Dr. Caroline McGill, who collected antiques and other memorabilia,

Hunting camp in the Gallatin. —U.S.D.A. Forest Service

found the door so interesting that she carted it off to the Museum of the Rockies in Bozeman.

A trapper named Kistner constructed another early, crude cabin in Levinski Park at the base of Lone Mountain. In the 1870s Kistner trapped marten on the slopes of the mountain, which he sold for as much as sixty dollars a skin.

As the century drew to a close, more and more Bozeman people came into the canyon to fill the stew pot. The hunters usually drove a wagon to Spanish Creek, and, as the county improved the track that served as a road, to Squaw Creek. There they loaded up their pack horses and picked their way carefully up the treacherous trail into the canyon. Slowly, the people in the Gallatin Valley learned of the bounty and the beauty of the area. Valley residents began to think of ways to exploit the resources of the canyon.

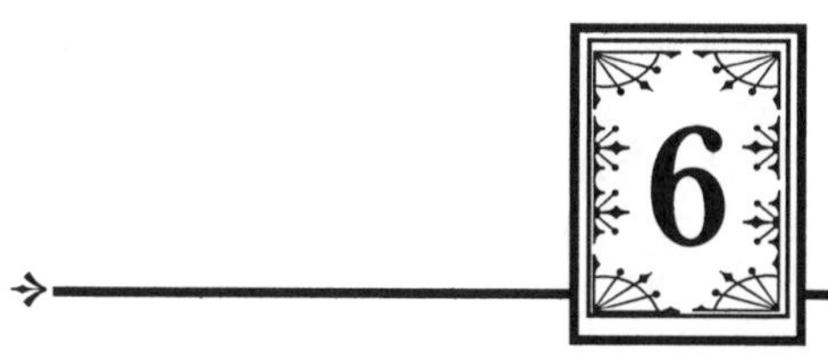

The Coming of the Railroad

In response to outside influences, the pace of life in the canyon began to quicken in the 1880s. Gallatin Valley farmers increased their grain production and raised more cattle to fill the needs of the mining towns in the Alder Gulch area. The growth of Bozeman and the region around it forced residents to go further into the surrounding mountains for the lumber they needed to expand their homes and businesses. Settlers from the newly formed town of Salesville (now Gallatin Gateway) penetrated the Gallatin Canyon as far as Spanish Creek. They used horses and sleds to skid logs out into the valley.

The approach of the Northern Pacific Railroad had the greatest impact on the canyon. By 1880 the railroad had reached the eastern boundary of the territory of Montana. Bozeman was the first settled town in the territory that the Northern Pacific reached on its march west, with settlement in eastern Montana following the railroad. Western Montana was already settled, and the residents were extracting gold, silver, and copper, raising wheat and cattle, and cutting timber before the anxiously awaited railroad arrived.

When the United States Congress chartered the Northern Pacific Railroad in 1864, it awarded the company every other section (640 acres) of land within a belt twenty miles wide on either side of the track in the existing states. In the territories, the federal government awarded the railroads every other section forty miles deep on either side of the track. The United States awarded the railroads odd-numbered sections and retained the even-numbered sections. The government also set aside an additional ten-mile-wide belt to provide the railroads with "lieu" lands. Often an odd numbered section within the forty-mile belt had already been claimed and was unavailable to the railroad. The company could then go to the ten-mile belt and get a section in lieu of the land denied it.

The railroad planned to sell its sections using the money to pay for construction of the line. They chose in lieu good agricultural lands they could readily sell. If agricultural lands were unavailable, the railroad chose heavily timbered land that would produce the ties and timbers needed for building tracks, trestles, and tunnels. The Northern Pacific received sections of land in the Gallatin Canyon as far south as Wapiti Creek in the Taylor Fork drainage.

In the early 1880s Zachary Sales, who had a saw mill at Salesville, set up tie-cutting camps for the Northern Pacific Railroad along the Gallatin River as far south as Taylor Fork. Because businesses faced a shortage of labor in the valley, Sales brought in most of the loggers for the job. These men left the canyon dotted with stumps, then moved on to other areas as the railroad made its way across Montana.

To surmount the problem of transporting the logs to the mill, the loggers dammed the creeks in the side drainages where they cut logs. They waited for high water in the spring, then rolled the timber into the stream. The loggers then broke the dam, and the high water floated the logs down the main Gallatin River to the valley. There sawmill operators took them from the river.

This method of transporting logs often ran into snags. The *Avant Courier* of 3 May 1883 recounted:

> A thrilling adventure and narrow escape on the West Gallatin. Williams and Company, tie contractors, have a force of men employed in floating the 'sleepers' down the river. . . . Persons familiar with the stream will recall how swift it is in the canyon. Sharp projecting rocks appear at intervals in the boiling waters. . . . There are no means of navigating the waters between the mouth of the canyon and the point where the ties were procured. . . .
>
> It became apparent when the ties did not come through, that they were lodged in the canyon. A man was sent along the path, and after a while returned to report a 'gorge' about midway between where the men were at work and the only apparently available spot where the stream could be reached below. The owners were at a loss but ordered the men to desist from putting other ties in the stream, in the hope that those lodged would loosen up and float out. Next day there was no change; not a tie had come through. An adventurous logger named Wilton proposed to go down and see if he could not start the ties. The owners would not consent, but finally seeing he was determined, told him that if he succeeded they would pay him handsomely but they preferred he would not make the attempt.
>
> Wilton, however, made the venture, but, to use his own words, there was not railroad companies or tie contractors possessed of

Rapids near Cave Creek where many log jams formed. —Dorothy Nile

sufficient wealth to induce him to again make a like attempt. When he started, he says, there was but little difficulty in proceeding the first mile. After that, progress became more difficult, and he entrusted his weight to a small bush, the shrub gave way and he was precipitated into the stream just where it seemed no human could ever come out alive.

By hard work, encumbered as he was with his clothing, he succeeded in keeping clear of the rocks, although borne down stream

> at a rate rapid and dangerous alike. After going some distance he thought to gain the bank but neglected to look ahead and came in contact with a snag, which tore and bruised him considerably. The cool waters aided to keep him in possession of his senses and after resting awhile he made another and successful effort to reach the bank. This was just about where the ties were lodged. He surveyed the manner in which they were fast and wisely concluding to take no more unavoidable risks, cautiously proceeded down stream, reaching the mouth of the canyon at dark—seven hours after he left the camp above. When he reached a place of safety he gave a look backward and was surprised to see the ties coming. He shouted and his call was answered by men on the lookout. He claimed credit for having loosened the timbers and accounted for his appearance with them saying he 'straddled a tie and picked out a road for them.' Although he told the contractors of his adventure and escape and disclaimed any credit for what he had not done, the gentlemen employers, in consideration of his good intentions, paid him a week's wages for his journey through the canyon.

It proved easier to float logs down the river than to supply the men who worked the tie camps. Walter Sales, in an autobiographical paper, tells about his first trip up the canyon in the middle of 1881:

> The trip was made with my cousin Henry. He was packing provisions to logging camps that were scattered all along the canyon from Sheep Rock to Taylor's Fork. The largest camp being at Twin Cabins where the Lemons now locate.
>
> Part of the way the trail was fierce and how a horse could ever get over it without breaking a leg seems like a miracle; but every once in a while one did break a leg and later years when they commenced moving cattle up the canyon for summer pasture, many of them did break their legs and fall off the trail into the river and this was the original reason for building the canyon road.

Sales made his trip during high water. This forced him to trail ten pack horses up Squaw Creek and around the mountain tops to Greek Creek, where he and his cousin once more returned to the trail along the Gallatin River.

As an alternative to using the trail from the mouth of the canyon, packers developed a trail that went up Middle Creek in the Hyalite Mountains and came down Squaw Creek. Travelers found either route difficult, and citizens lobbied the Bozeman County Commissioners for a road up the canyon.

Albert Greek, a Scottish immigrant, operated a sawmill at Greek Creek during the tie-cutting operation. Almost every foreign born person in the canyon had a romantic story associated with him. Albert Greek supposedly trained as a medical doctor in Edinburgh. He left

Albert Greek. —Brad Stratton

his practice and home in Scotland because of an unrequited love. A tall man with a trim beard that just covered his chin, Greek worked at the sawmill at Salesville for a while, then operated his own sawmill at the creek named after him. He helped construct the trail from Middle Creek to Squaw Creek to facilitate supplying his camp. Greek eventually sold his sawmill for $375 to Lewis Cass Bartholemew and disappeared from the Gallatin Canyon.

Farther up the canyon, just north of the future location of Karst Camp, Sam Harper of Bozeman built a log cabin used by lumberjacks. The camp became known as Robber's Roost because of the habit that the men developed. Each morning the first one up grabbed the best clothing hanging on the wall pegs, whether or not he owned it.

President U. S. Grant created Yellowstone National Park in Montana, Wyoming, and Idaho in 1872. Visitors traveled through the park by horseback and stagecoach. Bozeman businessmen looking for a way to take advantage of the nation's first national park, envisioned their town becoming a gateway city to the Yellowstone area: tourists would transfer from the Northern Pacific to a secondary line that would go through the Gallatin Basin to the northwest corner of the park. From the park boundary, the tourists would travel by stagecoach over Big Horn Pass, through Gardner Hole, and on to Mammoth Hot Springs. Besides bringing tourist revenues to Bozeman, the railroad would supply the lumber camps in the canyon and carry out coal and other minerals.

To lay the groundwork for this railroad spur, Walter Cooper and George Wakefield, Bozeman businessmen, and Peter Koch, a civil engineer, made an exploratory trip up the canyon in August 1881. Cooper later owned a large railroad tie-cutting operation in the upper Gallatin. Cooper and Wakefield checked for mineral deposits in the canyon, especially coal that could fuel steam engines. Koch surveyed a possible route for the rail line. Predictably, the men returned to Bozeman with enthusiastic accounts of the plausibility of a rail route up the canyon and coal deposits in abundance to fuel the trains.

The Northern Pacific Railroad decided against constructing a rail line up the narrow Gallatin Canyon due to high construction costs. Further, they would need to build a connecting wagon road through the northwest corner of the park to Mammoth Hot Springs. Instead, the railroad opted for the open and friendly terrain of the Yellowstone Valley to the east. This line led directly to Mammoth Hot Springs. So the Gallatin Canyon missed out on train-loads of tourists and also lost transportation for lumber, coal, and other minerals, one reason the canyon never developed a successful mining business.

Bozeman lost the railhead to the park, but the community kept up its push to gain easier access to the Gallatin Canyon. Concerned businessmen in Bozeman began to petition the county government for a road from Bozeman to the upper basin and the northwest corner of Yellowstone National Park.

The account that Bozeman businessman Walter Cooper wrote of his trip through the Gallatin shows that he assessed the value of the lumber in the canyon:

> August 19, 1881. Peter Koch, George Wakefield, Walter Cooper, with Nelson Collins, packer and Ford, cook, left Bozeman at 11 o'clock. . . . Made camp at mouth of Spanish Creek which has one-fourth flow of water as that of Gallatin. Caught three trout with brown hackle. . . . No sign of game so far. Canyon quite open and

opposite mouth of Spanish Creek wide, beautiful valley one-fourth to half a mile wide; hills and valley covered with a good growth of grass.

August 20th. Cook up at four o'clock. Breakfast at five. Left camp at 6:30. Crossed river mouth of Spanish Creek. . . . Entered Sow Belly Canyon one and one-half miles long; rough portion of trail over granite shale rock . . . fall of stream 80 to 100 feet per mile. . . . Scenery very fine from other side in Hell Roaring Creek Canyon. Limestone formation changes to granite. River here about fifty feet wide. On either side of the river cliffs rise to the height of one thousand feet to two thousand. Scenery grand. . . . Hills and mountains above Hell Roaring heavily timbered with spruce, pine timber suitable for ties for railroad. Timber grows in every bend of river in great abundance and should judge that from four to six ties could be made from each large fir tree. Still no sign of game. The angler would be in his glory.

August 25th. Cook up at usual time. Started for home at 6:30. Reached Lower Basin at 10. . . . Left to take direct course down the river Gallatin. Object was to ascertain character of country between mountains of Gallatin and Bozeman.

The 1881 burn in the canyon. —Museum of the Rockies Photo Archives

Of more interest, Cooper documented a large forest fire in the canyon that shows in photographs taken at the end of the century. On 1 September 1881 the *Avant Courier* reported the fire:

> Several days last week the volumes of smoke arising from there were so large and dense as to enshroud the entire valley in a pall of cloudy darkness. At midday, Old Sol appeared as if viewed through smoked glass, impressing the timid and superstitious with vague apprehensions of the near fulfillment of "Mother Shipton's prophecies." The greatest calamity realized, however, is the useless and serious destruction of large tracts of valuable timber. . . . The rain Tuesday probably extinguished the fires, but not before they had devastated thousands of acres of valuable forests. . . .

The fire started naturally on the Madison divide and winds fanned it through the Gallatin Canyon. Cooper's journal gives a vivid account of his experiences on the trip through the burning landscape along the Gallatin River:

> August 26—Up as usual. Traveled steadily. Water shallow in river. Fall 70 feet to the mile. Made twelve miles in forenoon. Halted at 1:45 for noon meal. Smoky and hazy ahead. Smell of smoke in the air. Push on till 6:30. Smoke denser. Must be approaching forest fire, probably not in our path. Seems advisable to rest for the night and push on early in the morning. Might be cut off if we go back.
>
> August 27—Up at 3:30. Quick breakfast. Got underway at 5:30. Keep to course of river - still shallow. Timber hard to break through. Smoke awful. Horses nervous. Birds with seared wings flying past us. No red glow anywhere but thick heavy smoke. Traveled all choked with smoke. Plenty of water, but must push on. Very short noon halt. Dark early and smoke worse. See red glow ahead and very nervous. Taking turns now sleeping. Afraid to travel before daylight for fear we might get off our course and get lost in the increasing smoke.
>
> August 28—For fear of what might lay ahead I'm trying to keep up journal. Day breaks late. Ready to start at 5:30. No use earlier. Fire creeping towards us from every direction; all around us. Horse's feet very sore. Traveled through hot ashes. Steer horses down creek bed whenever possible. River increasing in size as fed by streams down canyon. Water getting too deep to travel in stream. No grass or feed left for horses to crop except the very last of what we are carrying. Trying to keep Brownie [the Cooper dog, which had followed the party] riding on top of one of the packs. Her feet very raw. What this night will bring and tomorrow we know not.
>
> August 29—Make all possible haste. Trail easier to follow. Traveled greater part of the night. Flames jump canyon from one side to the other. Spectacular sight and an inferno. See small game with seared fur and bewildered elk and deer race by. We hardly halt for

food. Cannot travel any part of the way in stream. Hot ashes falling everywhere. Noise of fire deafening. By late afternoon reached burned over section and find heat not so intense. Stench of smoke still awful. Coughing constantly. Horses short winded and exhausted. We judge we have made fifty miles since two this morning. No feed left in the upper or lower part of the Gallatin Basin. Timber practically burned out. Only comfort we are getting well down the canyon toward home. Reached Hell Roaring at 7 o'clock tonight. Halting now for feed. One mountain completely bare of timber. We are naming it 'Baldy'. Fire out here. Some smouldering patches still and hot ashes everywhere. Some places the largest trees still standing. One wonders how they escaped. Will push on to Spanish Creek and camp near the mouth of the canyon late tonight.

August 30—Reached home at 8 o'clock P.M. Had good night's sleep at Spanish Creek last night. Still smokey but no further danger. Found hardly any feed for the horses but halted outside the canyon this morning and let them crop grass. Smoke still bad, valley seemed to be full. Found fire only confined to canyon. Probably started by lightening. Worst fire for many years. Horse's feet too sore to make any time. All day getting home. Will complete report to be sent to St. Paul.

This foot and horse bridge crossed the Gallatin River before construction of the new road and was made entirely of poles. —Museum of the Rockies Photo Archives

The Ten Foot Wide Road

As the cattle industry in Bozeman expanded, ranchers joined the timber interests in demanding a road into the canyon. The coming of the railroad promised a way to get beef to eastern markets. The Gallatin Canyon could provide much-needed grazing for cattle during the summer. Another group of homesteaders also wanted the road. Prospectors, who thought that the canyon would match other areas of the territory in mineral wealth, began to petition the county government for an easy way to haul their supplies into the basin.

Before Gallatin County existed, Nels Murray, a well-known trapper, had improved the Indian trail that went up the canyon. With the help of some friends, Murray constructed several miles of pack trail. Still, many unimproved miles remained.

The early trail from Bozeman met three fords near the mouth of the canyon. Travelers forded the river at the entrance to the canyon, below Spanish Creek, and at Spanish Creek. They then followed a poorly defined, meandering game trail that went along the high bluffs on the east side of the river. The trail passed below Sheep Rock and Castle Rock, then dipped down to Squaw Creek. After fording Squaw Creek, the track stayed at river level to Cave Creek.

Following a treacherous ford at Cave Creek, travelers proceeded along the west side of the river for a few miles. They then crossed back to the east side via yet another ford, and faced one more ford before finally reaching the lower basin. From the basin to Taylor Fork the trail followed along the west side of the Gallatin River. Little wonder that valley residents who wanted to use the canyon urged the county commissioners to construct a road and bridges.

In January of 1883 the *Avant Courier* of Bozeman described the route into the canyon as "a dizzy bridle path [that] leads along the side of the mountain. The range rises abruptly some 1,500 feet above the path. An equal distance below flow the waters—boiling, seething—madly dashing in their course to the Gulf."

The trail made supplying the early lumber camps a problem, as freighters had to pack all their supplies over it. Cattlemen faced an even more difficult task. They had to drive their cattle single file over the treacherous trail. It took several days to move a herd of cattle from the valley to the basin.

In 1883 the Gallatin County Commissioners sent Engineer Bundock to survey a wagon route through the canyon to Yellowstone Park. As reported in the *Avant Courier*, Mr. Bundock found a good wagon road from Salesville to Squaw Creek:

> Fourteen miles above Squaw Creek it will be necessary to cross the river. The heaviest part of the work is found between Squaw and Swan Creeks, there being seven miles of bluff work in the distance.
>
> The Gallatin Basin is entered forty-one miles from Bozeman. There is little work from there to the Second Canyon—seven miles—and less work after the canyon is entered.
>
> Mr. Bundock further said he regarded the route as not only inexpensive in construction, but as one that should long since have been opened up to tourist travel. It will open up to settlement and cultivation vast acres of fertile lands. The timber growth along the water courses is unusually fine. There are fine coal beds, extensive mineral tracts, abundance of game and delightful scenery. We regard the road as one that should long since have been built, and one that is now imperatively demanded.

Sixteen years of consistent lobbying followed before the county approved the construction of the road to the upper basin.

Bozeman resident W. W. Wylie for many years operated tourist camps and provided transportation through Yellowstone Park. At Spanish Creek Wylie grazed up to three thousand horses that he used in his tourist operations. After taking a trip up the Gallatin Canyon in 1887, he wrote, "I left Bozeman with a small party on horseback and with pack mules, to enter the park at what is now the western entrance, or the town of West Yellowstone. No town was there then. We traveled the Gallatin River from the Gallatin Valley. There was at that time no trail along the river over which we could ride without frequently having to cut fallen trees out of our way so that our saddle and pack animals could pass."

The canyon received continually increasing use, but still lacked a wagon road. The *Avant Courier* in August of 1889 reported that "Charles Anceney has moved his cattle to the Gallatin Basin. He found it difficult work trailing them through the canyon and lost four head. He is very well pleased with the new range."

Bozeman businessmen continued to realize the economic importance of the canyon and continued to press for a road. The *Avant Courier* in September 1890 reported:

> A large number of the citizens of Bozeman and surrounding country are taking a deep interest in and urging the importance of building a wagon road from the Gallatin Valley, proper, to the upper Gallatin Basin. A numerously signed petition was presented to the board of county commissioners, at its recent session, praying for the establishment of a public highway to the point named. . . . and that, as the proposed road would lead by a direct and short route to the upper Geyser Basin, it would be extensively traveled in summer time by tourists, visiting the Yellowstone National Park by private conveyance. . . .
>
> These are all matters worthy of careful consideration by the county commissioners, as well as by our citizens generally, inasmuch as their availability and acquisition would importantly add to our present population, considerably increase the assessable property of the county, materially advance the commercial interests of Bozeman and otherwise contribute to the general prosperity of the entire community.

In October of 1890, the *Avant Courier* reported an estimated cost of the proposed road through the Gallatin Canyon to the Park at $10,000, "a good investment for Bozeman and Gallatin county."

In November the paper enthusiastically suggested that instead of making all county residents pay for the road, the residents of Bozeman, who would benefit most, should be willing to pick up most of the $10,000 tab, " . . . a remarkably small amount in comparison to the benefits which it is believed would accrue to the entire county." In January of 1891 the county commissioners denied the petition for the hoped-for road.

In March 1892 Lewis Michener, the father of Thomas Michener, who left his unfinished history of the canyon, presented the county commissioners with a petition signed by six hundred residents. The petition asked for the construction of a road from Bozeman to the Yellowstone Park line via the Gallatin Canyon. Michener, a resident of Salesville, had spent five years riding around the county gathering names on the petition. At that time only men who had paid their poll tax could vote, so it took a lot of doing to get six hundred names. Michener wanted a wagon road so he could easily reach the many mining claims he had filed on in the Gallatin Canyon. The *Avant Courier* stated that the "said petition was given a first hearing and found to be in accordance with the

A buckboard driving up the canyon on the new road. Note the burned trees.

law, so the commissioners appointed reviewers to examine the route of the proposed road and file a report thereof."

During the next few months, other business interests joined Michener in urging construction of a road up the Gallatin. An item in the *Avant Courier* of 22 January 1892 says:

> The advocates of the West Gallatin Road came before the board. J. G. Lane was prominent among the respectable members and citizens who appeared before the Board of County Commissioners on Saturday last in advocacy of the West Gallatin wagon road to the basin. Jack Lane held the floor for half an hour while in true Patrick Henry style he demonstrated . . . the importance of the enterprise, reminding the County Commissioners that they were servants and not masters of the people. He called attention to the petition before the board signed by

prominent tax payers of the county, praying for a suitable appropriation and such favorable official action as would insure the speedy construction of the highway and against which there was not a single vote. Mr. Lane was repeatedly applauded by the delegation of private citizens present during the delivery of his earnest and eloquent speech. L. Michener and others also ably urged upon the board favorable action in the matter.

In May of that same year representatives of the county commissioners set off on a trip to view the route of the proposed road. The *Avant Courier* reported that: "L. Michener and party of Gallatin basin road viewers returned to Bozeman Thursday afternoon. They had fair weather during their trip of thirteen days and they encountered less snow than they had expected. From their observations they feel certain that a road can and will be established from Bozeman to the Park." Lewis Michener received $57.50 for furnishing horses and camp equipment for the group that made the trip. In May the commissioners authorized a payment of $750 to have the Gallatin Canyon route surveyed and to arrive at a cost of construction.

In September 1892 county surveyor C. M. Thorpe completed a survey of a proposed canyon road and filed his report and map. On 14 December the Commissioners ordered that the clerk insert in the *Weekly Chronicle* "for 30 days an advertisement for bids for building the West Gallatin Road and bridges."

The grade on the new road near Portal Creek.

At the commissioners' meeting in February 1893 the county attorney advised that the county could not legally incur the expense of building the road without submitting the matter to the voters of the county in the manner proscribed by law. More false starts were made in the following years.

For the next five years interested residents petitioned the board to build the road up the canyon. Lewis Michener, who had collected the six hundred names on the petition for the road, died. His son Thomas Michener took up the cause. Homesteaders like Thomas Lemon, who lived at Spanish Creek, and prospectors, like Thomas Michener, had a personal interest in reaching their property in the canyon in a reasonable manner. Charles Anceney, Sr., and Still Huling favored the road so they could quickly and safely drive their herds of cattle to summer pasture. W. W. Wylie wanted the road so he could move his horses from Yellowstone National Park to winter pasture at Spanish Creek. Walter Cooper wanted the road built so he could easily and economically supply the many tie camps that he planned to set up in the Gallatin Canyon. This diverse group of people, some influential, some just ordinary citizens, petitioned and prayed and lobbied. Their pleas fell on deaf ears.

In March 1897, Mr. Cockerell, a contractor and businessman, petitioned the county for $200 to build two bridges. He planned to put one bridge over the West Fork of the Gallatin River (at the entrance to what now is Big Sky) and the other about one mile below the West Fork. The board took Cockerell's petition under advisement, finally allotting money to build the bridges.

Now that two bridges would soon span the Gallatin River in the basin, interested parties petitioned the commissioners to complete the road to the proposed bridges. As word about the roadless bridges spread, the commissioners called for bids to finish the road. At a special meeting the commissioners received the bids from: V. A. Cockerell ($8,098.50), M. J. Brow ($8,766.8), and Bradley and Dean ($8,195.70).

Mr. Cockerell won the contract for constructing thirty-two miles of road starting at Squaw Creek and ending at the park line. He was to finish the work by July of 1898. Bad weather caused delays and Cockerell completed the road, which measured ten feet wide, and the bridges in late August 1898. The work stopped at Taylor Fork, four miles short of the park line. The county conducted negotiations with the U. S. Government for extending the road though the park.

In July 1898 the first wagon traveled up the canyon road. Lumberjacks had brought dismantled wagons into the tie camps at Taylor Fork, but no one had driven a wagon up the canyon. Tom Michener, anxious to get into his place at the West Fork, headed up the new road with his mother, younger sister Myrtle, and his bride-to-be, Mary Elizabeth Lockhart. Mother and daughter rode in a buckboard. Miss Lockhart,

Mamie and Tom Michener

wearing a wide divided skirt, rode astride her saddle horse. The group found the bridges at Cascade Creek and Cave Creek unfinished, and the horses were unable to cross over the large rocks at Cave Creek. So Miss Lockhart took the team up around the cliffs and back down to the river where she met the others at an easy ford. It pleased Tom Michener to have everyone see that his bride-to-be was a fine horsewoman.

The men were just finishing the bridge below the West Fork. Miss Lockhart forded the river with the team, and the workmen laid poles across the frame of the bridge so they could push the wagon across. The new bridge was close to the water and the workmen jumped in to help the wagon along. One of these men wore a buckskin suit that shrunk in the water, with the cuffs of his pants ending up around his knees. It must have been uncomfortable, but it provided the other workers with a bit of merriment.

Horses and buckboard reunited and continued on the road to the West Fork bridge, which workers were also finishing as they arrived. The horses forded the river again, and the workmen laid down poles and once more pushed the wagon across the bridge. The men, most of whom lived in the basin, threw their hats in the air and cheered because they had waited so many years for this moment. With the completion of the road from Bozeman to the upper basin homesteaders and entrepreneurs alike would find their everyday life in the canyon easier.

Once the road reached Taylor Fork, canyon residents spent ten years trying to get a road into Yellowstone Park, preferably to Mammoth Hot Springs via the Big Horn Pass. From the Gallatin River to Mammoth the route followed part of the Bannock Indian Trail. Indian trails usually traversed south facing slopes where the snow melted early in the spring. A gentle grade led from the Gallatin to Mammoth. Canyon residents felt that this route would have entailed less work than any other way into

Michener Camp about 1902.

the park. Tom Michener took some county officials and drove a buggy through Big Horn Pass to Mammoth to show that they could easily construct a road along the trail. Michener wrote articles for various magazines to try to influence park officials. In 1904 a committee of Bozeman citizens wrote U. S. Senator Paris Gibson asking him to press for construction of the road from the Gallatin River into Yellowstone by way of Big Horn Pass.

The U. S. Park Service surveyed the Big Horn Pass route in 1907 and then pigeonholed the proposal, probably for two reasons: first, Lt. Ernest Peeks, who did the survey, estimated that a crude road would cost more than $32,000, and a more adequate road would cost $100,000; second, the Union Pacific Railroad had just built a feeder line to the present site of West Yellowstone. Although the town had yet to be established it seemed certain that one would spring up at the railhead. It made sense to build the road from the Gallatin Canyon to the west entrance to the park by way of Grayling Creek.

When the officials were deciding which route to use, many people bet on the decision. Sam Wilson had traveled over the Big Horn Pass Trail into Mammoth and felt it was the only sensible route. So he bet one of the county commissioners a Stetson hat, even then an expensive wager, that the road would go over the pass to Mammoth instead of along Greyling Creek to West Yellowstone. Sam Wilson had to come up with a new Stetson when the road into Yellowstone Park went south instead of east.

Part of the road went through the northwest section of Yellowstone Park, making it necessary for Gallatin County to get permission from the Park Service to run the road south. In May 1910 the Gallatin county commissioners obtained approval from the U. S. secretary of the interior to survey and build a road through the edge of the park. This road cut by one-half the distance between Bozeman and West Yellowstone. The growing number of automobiles had convinced park officials of the need for the road.

Building the road down Greyling Creek proved more difficult than crossing Big Horn Pass. The road ran through many marshy areas. Construction involved almost fifty culverts and pole bridges, and workers had to lay down a corduroy road of logs to keep vehicles from sinking into the mud. The hard grade was difficult to keep open, and for many years the road was impassable in the winter.

The new road opened the canyon to exploitation. Loggers' wagons soon rumbled over the road with provisions for the tie hackers who stripped many areas of their trees. Cowboys pushed herds of cattle up the road to graze on the grass in the basin. Prospectors drove wagon loads of supplies into the area and scoured the land for minerals. And, making the most significant changes to the canyon, homesteaders squatted on land and found ways to make a living.

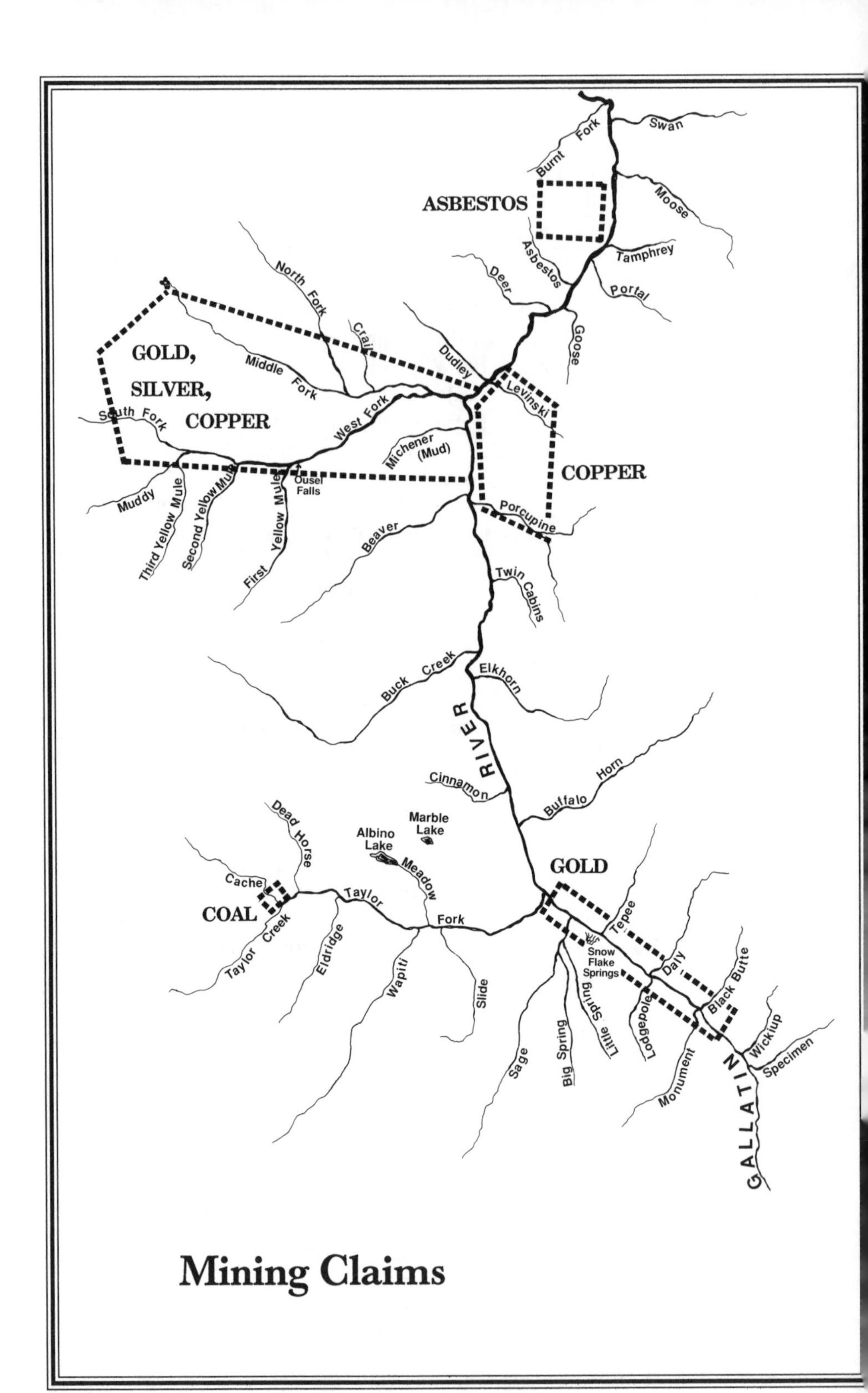

Mining Claims

Early Mining Claims

Soon after lumberjacks finished cutting timber in the canyon in the early 1880s, Bozeman and Salesville individuals began pasturing horses and cattle at Spanish Creek during the summer. As homesteaders took up the free grasslands, large cattle interests pushed on to the basin. Trailing several hundred cattle single file over the steep, rocky trail took several days. Sometimes cattle fell from the cliffs high above the Gallatin River and washed downstream. Yet the free range more than made up for the loss of a few head.

Charles Anceney, Sr., moved cattle into the basin in 1889. In 1894 Still Huling, joined by his wife—who dressed like a man and herded with the best of the boys—pastured their cattle in the upper basin. The cattle ranged over Tepee Creek and Daly Creek as far south as Black Butte. The Hulings had a cabin in Tepee Creek.

The herders usually stopped several times on their way up the canyon to spend the night. They looked for flat spots where the cattle could graze and bed down without being in danger of falling off a cliff and made the lodge of the Duke of Hell Roaring one of the their favorite stopping spots.

A legend has grown up around the Duke and his wife, reputing him to be an Austrian count in exile, and she to be a Russian-born aristocrat. They settled on Hell Roaring Creek and were known as Mr. and Mrs. Wagner, although people usually called Mr. Wagner the Duke of Hell Roaring.

Wagner built their large log cabin and decorated the walls with his collection of guns and swords. The Duke made a living hunting and trapping. Mrs. Wagner served meals to herders, lumberjacks, trappers, and any other traveler who ventured up the canyon, earning a good sum

Still Huling and Mrs. Huling.

of money for her work. Legend has it that Mrs. Wagner wore the delicately embroidered skirts of the Russian gentry, smoked cigarettes, and drank black coffee.

The Duke's hunting brought in little money, and circumstances forced the Wagners to sell their place in the canyon. Mr. Dier, owner of the Bozeman Hotel, bought it for a summer home. Mr. Dier tells of visiting with the Duke for a few days at his canyon home to negotiate the purchase of the place. The Duke had a quick and viscous temper. During the negotiations the Duke grabbed a pistol, held it to his own head, and exclaimed, "Sometimes I get so discouraged I could easily do zees." He gave his head a viscous poke and then calmly placed the gun on the table while Mr. Dier almost had a heart attack.

Shortly after Mr. Dier bought the land, the Duke of Hell Roaring and his wife disappeared from this part of the country. The legend surrounding the couple reports that Mrs. Wagner left the duke and returned to her home in Russia. We know more conclusively that the Duke went to Yellowstone Park where he trapped and hunted and did other work for park officials.

Even before construction of the road, many prospectors drifted into the canyon, and they often herded cattle. Labor was scarce, and prospectors wishing to spend a pleasant summer could make enough money herding to finance their search for gold and other minerals. Herding required only periodic checking of the cattle, which gave prospectors plenty of time to pan the streams. John Wylie of Bozeman ran cattle in the canyon and Thomas Michener and William Lytle, both of whom homesteaded in the canyon, herded for him. Michener and Lytle ran Wylie's cattle in the Porcupine drainage and on West Fork, where Michener later filed many mining claims.

Pleun Robol herded cattle for several people including Attanas Vioux, Ed Blackwood, and Patrick Corcoran, all of whom lived in the Gallatin Valley. Robol usually took care of four hundred head. (A cow with a calf at her side counts as one unit.) Like many other herders Robol eventually homesteaded in the canyon.

The lower canyon saw limited mining activity. Prospectors filed a few mining claims in the Spanish Creek area in 1872, about eight years after settlers moved into the Gallatin Valley. Mineral seekers filed ten

Pleun Robol.
—Doris Peterson

additional claims in the Spanish Creek area in the remaining years of the nineteenth century.

Compared to Spanish Creek, the upper canyon became a whirlwind of mining activity. In October 1886 Andrew Levinski filed the first claim in the upper canyon. The Livingstone Claim for gold, silver, and other minerals—one of many claims filed by Levinski in the basin—was located on unsurveyed public land, "between the head waters of the West Branch [Fork] of the Gallatin River and the head of Jack Creek on the divide between the lower basin of the Gallatin River and the Madison Valley."

In 1887 Josiah Pinkerton, Levinski's trapping partner, filed on copper, silver, and gold. The claim was located two and one-half miles north of Porcupine Creek on the east side of the Gallatin River. Levinski and Charles Carson filed three other claims in the same year and in the same general location for copper and other minerals. By 1889 Lewis and Thomas Michener had filed a claim to gold, silver, and other minerals, giving the location as "one-half mile north of a place known as the Burned Cabin and about one thousand yards west of the Yellowstone National Park line."

Prospectors filed about thirty-five claims in 1890, thirty in 1891, twenty-four in 1892 and fifteen the next year. Miners filed thirteen more claims before the end of the century. The claims were grouped in four areas: west from the mouth of the West Fork to Lone Mountain and Jack Creek; along the east side of the Gallatin River from a point below Levinski Creek south to Porcupine Creek; in the upper basin near Black Butte; and near Karst Camp.

Prospectors in the canyon panned for gold in streams and examined quartz veins for evidence of precious metals. If prospectors found even a trace of a valuable mineral, they could stake a claim on a twenty-acre parcel of land. Claims usually followed a vein or a river.

The act of claiming required prospectors to drive a stake in the ground or to build a stone monument at the point of discovery and to affix to the stake or monument information including the name of the claimant, minerals claimed, physical description of the claim, and date. Additional stakes or monuments marked the four corners of the claim. Prospectors had to do this for each twenty acre section claimed. Then the prospectors would hurry to Bozeman to record their claims with the county government. The county issued prospectors notarized claim papers that they attached to the stakes or monuments. The prospectors usually put the claim papers in tobacco cans, placed upside down to keep rain from seeping in. It was illegal to remove cans with paper in them, but anyone could take the papers out of the cans and read them. Once prospectors claimed land they had to perform $100 worth of work a year on their claims. The law required prospectors to record proof of the work for the preceding year by 1 January.

Before 1906 prospectors in the canyon sought minerals on unsurveyed public lands. After the formation of the Gallatin National Forest, prospectors usually looked for minerals on national forest land. In either case prospectors usually gave original and inexact descriptions of the land. The prospectors owned the claimed minerals, and the land belonged to the United States. Theoretically, prospectors could lay claim to the minerals on homesteads, but this never happened in the canyon. Prospectors could also strike deals with the Northern Pacific Railroad to extract minerals on its land, paying the railroad a royalty.

The early prospectors all looked for gold—prospectors have never found silver worth mentioning in the Gallatin Canyon and the canyon has too little iron to be of commercial interest. The massive copper

Pat Shane at West Fork.
—Museum of the Rockies Photo Archives

deposits found at Butte and Anaconda overshadow the small amounts of copper found in the Gallatin Canyon.

Early explorers found anthracite coal in Taylor Fork and in Coal Tunnel Mountain just south of the West Fork. The coal burned hot enough for use in blacksmithing shops, but since canyon residents had no way to transport the coal out of the canyon, it had no commercial value.

Some claimants lived in the canyon for many years, homesteading and contributing to the community. A few men recorded their claims in Bozeman, realized they were worthless, and moved on. Over the years the Michener family filed forty-five claims. Pete Karst filed thirty-two claims. Pat Shane, grubstaked by numerous individuals, prospected for many years. He must have been illiterate because he signed his thirty-two mining claims with "his X mark." The claims are listed under Shane, Schein, and Sheehan. Andrew Levinski filed twenty-four claims. Charles Anceney, Jr., Hugh Beatty, Lewis Bartholemew, O. P. Chisholm, Y. J. Kyunders and A. Z. Stillman all filed claims from 1890 through the early years of the twentieth century.

A few prospectors and grazers used the canyon for short periods and then sold out their squatters' rights. Those who stayed were a colorful bunch: prospectors, herders, sawyers, builders, hunters, lumberjacks, and trappers. Lewis Michener, the man who collected signatures for a road up the canyon, had been a butcher at Alder Gulch and had come into the canyon in the early 1870s hunting elk for his butcher shop. He had found a copious sweet spring at the mouth of the West Fork and lush meadows in the West Fork drainage. In 1880 he ran a herd of horses over Jack Creek, summering them on what is today the golf course at Big Sky. The Micheners homesteaded at Meadow Creek in the Madison Valley. Construction of the dam that created Ennis Lake forced the Micheners to sell their property to the power company. They moved to Salesville but continued to pasture horses in the canyon. They also captured deer and elk calves, which they raised and sold to the Columbia Gardens Zoo in Butte and the zoo in Spokane, Washington.

By 1890 the Michener family had a cabin at West Fork, near the spring that Lewis had discovered. Each summer Lewis and Emmaline Michener and their seven children piled all of their supplies into a wagon and drove from Salesville to Spanish Creek. At Spanish Creek they transferred to horses, putting their belongings on pack horses. Mrs. Michener and the girls rode sidesaddle with Mrs. Michener always carrying her cat and her canary into the canyon. (She needed the cat to control the mouse population in the cabin at West Fork.) The Micheners brought in a cook stove and a wagon by dismantling them and packing them over the Madison divide on the Jack Creek Trail, an easier route than the one up the canyon.

Lewis Michener spent his summers prospecting for gold. Because they had already homesteaded, the law prohibited the Micheners from filing on another homestead but their son Thomas filed on the property his father had discovered. Both Micheners filed on mining claims.

Lewis and Thomas Michener, in partnership with Philo Tomilson and Lewis Cass (Big Lew) Bartholemew, started the second gold rush in the Gallatin Canyon about 1893. The four men found gold in the Black Butte area. Big Lew and Tom Michener stayed in the upper basin to hunt game. Lewis Michener and Tomilson went to Bozeman to buy supplies and file claims. Tomilson later said that the only mistake he made was telling Walter Sales about their find, which resulted in gold seekers rushing up the Gallatin to get in on the big strike. Tom and Big Lew saw a crowd of prospectors as they neared camp with fresh meat, so they faded back into the woods. Lewis Michener and Tomilson returned up the canyon as far as the West Fork where they ran into a crowd. Michener and Tomilson met with representatives of the group and received their promise to respect the claims of the original four. None of the stampeders found gold, and they went back to Bozeman a few days after they had arrived.

In 1917 Thomas Michener wrote first-hand biographical information on some of the men who prospected in the canyon:

> Jim Stewart was one of the old time cow punchers. He was born in Utah in 1859 and followed the cow trail from the time he was a small boy. He punched cows when the Indians were on the war path, when cattle were as wild as buffalo, when there were no railways, telephones or telegraphs and when the fastest messages were carried on the backs of a cayuse. He quit the cow business to hunt and prospect for a few years in the upper Gallatin. He finally settled down to a few horses and cows on a ranch somewhere close to Dillon.
>
> Hugh Beatty was an old lumber jack from Michigan. He had trouble with some of his wife's relations and had to jump the county two days ahead of the sheriff. He came to the Gallatin and lived almost the life of a hermit on some claims. About his only accomplishment was his wonderful proficiency in swearing. He was surely an artist along those lines.
>
> Andrew Levinski was Prussian by birth but of Polish parents. Very little is known of his early life. He came to Syracuse, New York, when just a boy . . . and finally landed on Spanish Creek where he lived with the Josiah Pinkerton family. He and Pinkerton trapped for a while on Spanish Creek and then moved further up the river [Gallatin] and Levinski . . . took up his abode in a cabin called Robber's Roost. . . . Game was plentiful, especially elk and deer and Levinski lived on meat. He would not work and the only money he got he secured through selling a few pelts or elk tongues. Levinski had shacks and lean-tos all over the canyon . . . he did build a cabin on Levinski Creek. . . .

Jack Smith is an old confederate soldier who in 1872 with Charles Blakely was perhaps the first to go through Buck Creek Canyon. He has been a cow man on the river since cattle have been in the basin." [Smith and Blakley were coming from the Madison Valley, following an Indian trail or a game trail and by accident picked Buck Creek Canyon, very narrow and heavily wooded, to descend. They must have had an extremely difficult time reaching the Gallatin River.]

Bert Stillman, one of the present citizens of Salesville, came to the canyon in the company of L. C. Bartholemew in the winter of '89. They trapped the canyon that winter and moved on to the basin in the following spring. They have both been associated, more or less, with the affairs of this place ever since.

L. C. Bartholemew has now changed his name to L. C. Bart, but he is better known as Big Lew. He is one of the unique characters of the mountains. He is a big, raw boned man, over six feet and weighing about 200 pounds. He is very proud of his large mustache. When I first knew him he used to dress in buck skin and with his knives and tomahawks he would have been a fine illustration as the hero of a dime novel. He is thoroughly American. I think he was born in Ohio but lived most of his early life in Missouri, Kansas and Texas. He always wants to do things differently from anyone else. He turns a grind stone in the opposite direction. He saws off the muzzles of his guns and changes the stocks just to be different. He uses a sundial instead of a clock. He has a good education and always pushes for reforms. He is an ardent advocate of socialism but worships at the shrine of a man with money. I have taken some time to describe Big Lew and hope that I am not doing him an injustice but rather apologizing for some of his infirmities.

Jack Smith at his cabin one mile north of West Fork. —L. Lawrence

> Captain Robbins was an old bear trapper on the Gallatin. He was a man that never had one day of schooling but with his wife's assistance he educated himself and taught grammar school when not engaged in bear trapping. He was a Captain in the Union Army during the Civil War and was killed at Butte in the big explosion of '94.

Red Wetzel had mining claims below the mouth of the West Fork. Wetzel had come west as a cook with General Reno. Many different people grubstaked Wetzel, which allowed him to spend his time prospecting. At the time Wetzel made his squatter's claim the government had not yet completed a survey of the land, so he had to mark his boundaries by metes and bounds. Wetzel, unable to remember these terms, referred to them as "leaps and jumps."

Lew Bart (Big Lew) and Bert Stillman located farther down the canyon. In 1889 these two men spent the year at Greek Creek, where Bart bought Albert Greek's sawmill. Bart killed thirty-five deer for market that winter. Later Bart squatted farther up the canyon. Bart sold his squatter's rights to Charles Oliver for $375. Oliver eventually sold to Pete Karst. Bart then continued up the canyon to Twin Cabins Creek. In addition to prospecting, Bart hunted, trapped, and ran his sawmill. In 1898 Bart enlisted in the Spanish American War and spent one and one-half years in the Army before returning to the canyon in 1900.

Ora Michener Lemon remembers Y. J. Kyunders, a Hollander who lived near the Michener family. Kyunders made a trip to town one day to buy supplies and returned packing an entire set of encyclopedias on his horse. When asked why he bought them, he replied, "If I should happen to forget something I can look it up." Mr. Kyunders never homesteaded, although he had a cabin at West Fork. He filed at least twelve mining claims over the years that he spent in the canyon.

Most people used the canyon in the summer but didn't spend the winter there. In January of 1890 Walter Sales, as an agent of the United States government, rode his horse into the basin to take the census. He found only two men, Andrew Levinski and Pat Shane, both of whom trapped beaver and prospected. By the turn of the century a few dozen prospectors had taken up squatters' rights in the canyon.

Michener Family, Tom and Mamie with Maggie, Ora and Charley, 1904.

The First Settlers

After 1898 a little community grew up around the junction of the West Fork and the Gallatin River. Mamie and Tom Michener and their growing family located at the spring known so long to their family. Pat Shane, Y. J. Kyunders, Pearl Lockhart, Red Wetzel, Pleun and Ida Robol, and the Frank Blanchard family either homesteaded or had mining claims in the area.

Pat Shane's homestead adjoined the Michener property. The numerous people who grubstaked Pat include Nelson Story, Jr., Bozeman businessman and political figure. Shane deeded his property to Nelson "Bud" Story III, who took care of Pat in his old age. The original Shane homestead is the site of the Soldiers Chapel, erected to the memory of Nelson Story IV, who died in World War II. The irascible Shane argued with all of his neighbors. One day Shane and Bert Stillman had a fight over how to make beans. Stillman had to hide Shane's rifle and put Shane's crutch on the roof to get away safely.

Pearl Lockhart, a sister of Mamie Michener, took up a homestead just below the West Fork. She donated her cabin for use as a school the summer of 1908, when the first teacher came into the canyon. Pearl later married Billy Lytle, and they lived south of Porcupine Creek. She never proved up on her original claim below the West Fork.

Pleun and Ida Robol, a sister of Tom Michener, took up a homestead on Porcupine Creek but did not stay to prove up. Soon after the birth of their daughter, they left the canyon for Reed Point where they ranched. Pleun said he didn't mind snow coming up to his waist, but when it started blowing in the back of his shirt collar it was time to move to another area.

Y. J. Kyunders never homesteaded but had a mining claim at West Fork. The Michener children remember that he ate a lot of meals with the family. Red Wetzel also had a mining claim in the area. Frank Blanchard and his family bought out the Dudley homestead claim and proved up on the Dudley Creek property. Frank ran a sawmill and did logging operations up and down the canyon.

Another prospector, Andrew Levinski, homesteaded on land next to the stream named for him. In addition, Levinski had cabins scattered through the rocky hills between his homestead and Porcupine Creek. The reclusive prospector lived on wild game. He shunned company, except the few valley residents whom he guided on hunting trips. Charles Anceney, Jr., recalled his first meeting with Levinski. Anceney found Levinski stained brown as a walnut by wood smoke. He wore the hide of an elk, with his feet in the hocks, which served as moccasins, and with the rest of the hide tied around him like a cape.

Up the West Fork drainage the Franklin Augustus (Frank) Crail family put together a 960-acre ranch, which today is Meadow Village at Big Sky. Frank Crail arrived in Virginia City in 1864. He carried freight from Fort Benton to Salt Lake City until trouble with the Indians drove him to the Gallatin Valley. He developed a strain of wheat, named Crail fife, which valley ranchers grew for many years. Frank Crail served as

Crail homestead: Lillian, Frank and Sally Crail with Mr. Creek, Sally Crail's father. —Elaine Hume

Gallatin County commissioner from 1896 to 1900. In 1902 he came into the Gallatin Canyon and liked what he saw.

Frank and Sally Crail proved up on their homestead and later bought 160 acres each from Sid Frazier, Robert L. Inabnit, and Ed Sprague. Their two children, Eugene and Emmet, then homesteaded, and the Crail family amassed 960 acres. When Crail first came into the canyon he ran horses and later raised cattle. The family had a large hay-cutting operation in the big meadows of the drainage. They put up hay, stacking it with a beaver slide that is still in use in the Big Hole Valley.

In 1915, in an attempt to make more money, Frank started running sheep in the West Fork drainage. Frank died in 1924, and Emmet continued to run sheep until 1934, when he went back to raising cattle. Neighbors often wondered how Emmet ever got his work done because he moved so slowly—one mild year Emmet put up the last load of hay on Christmas Eve. In the late 1940s Emmet married his sweetheart of twenty-five years, Anna Brennaman, and the neighbors gave the couple a large shivaree.

Eugene Crail served in the Army in the First World War, where he learned carpentry. He returned to the canyon and built the first Ophir School and the chapel at Rockhaven. Eugene did the log work on the cabins at the Lytle Ranch, later known as the B Bar K and today as the Lone Mountain Ranch. He built the Crail house, which is preserved in Meadow Village. Winters, when the weather prevented him from doing outside work, he made violins. The Crail's daughter, Lillian, trained as a nurse in Chicago and did not return to live in the canyon.

Clarence Lytle homesteaded the land now occupied by the Lone Mountain Ranch. He ran cattle and cut hay. Clarence often lived with his older brother Billy and leased his land out for grazing. Clarence served on the school board, and during the First World War he served on the draft board.

Three other homesteaders lived in the drainage. The Henry Johnson family located at West Fork Meadows. Mr. Johnson cooked at various dude ranches, hunting camps, and cattle outfits in the canyon. Cash Burnett, a young man from South Carolina, homesteaded up the Ousel Falls Road. He left to fight in World War I. After the war he returned to the canyon, sold his property to Lewis Michel, and left the canyon for good. In 1910 the Joseph Davis family homesteaded on the land at Hidden Village. Joe's son Charley proved up. The Davises, like many other homesteaders, only lived in the canyon during the summer.

"Happy" Jack Griffin, trapper, hunter and prospector, had a cabin on the West Fork. Griffin neither homesteaded nor filed a mining claim, despite being one of the earliest users of the canyon.

At the top of the divide between the Gallatin and Madison Valleys, a Civil War deserter named Jake Ulery dug ditches and lakes in a scheme to divert water from the Gallatin drainage and sell it in the drier Madison Valley. Around the turn of the century Ulery channeled water from Beehive Basin to a series of manmade lakes that now bear his name. The Gallatin County water commission learned of Ulery's scheme and stopped him from completing his plan. You can visit the three lakes north of the Mountain Village.

Jack Smith located at the first bridge below the West Fork, still known as the Jack Smith Bridge. Andrew Jackson Smith came from Missouri

Charley Smith at his cabin. —L. Lawrence

and first ventured into the canyon in 1872 on a hunting trip. In later years he pastured cattle for valley cattlemen Vard Cockerell and Ed Blackwood in Porcupine. Smith constructed corrals on the cliffs on the west side of the river to keep the cows from sliding down the steep slope into the Gallatin. Jack also rounded up strays and sold them back to their owners at $5 a head. In the early 1900s Jack worked for many years as a deputy game warden.

Jack's son Charley Smith made his first trip into the canyon in 1892 to hunt. The following year he brought cattle in for the summer. Smith resisted the urge to prospect and spent his summers guiding hunters and geology students from Harvard University through the canyon. Charley prided himself on teaching the professors and students to be self-sufficient. His rule in camp was that each person cleaned his own game and his own fish. One professor refused to clean his fish, so Charley cooked it just the way the professor handed it to him. He never had trouble with the Harvard group again.

Smith homesteaded north of Beaver Creek on the west side of the Gallatin River, where Buck and Helen Knight now live. His cabin still sits on its original site. The homestead has a clear spring, flat land for haying, and aspen-framed views of the cliffs along the east side of the river. Smith irrigated his hay with water from Beaver Creek and the irrigation ditch is still in working order. George Lemon, who has lived for seventy-two years in the canyon and used to help Charley put up hay, remembers Charley as a fastidious housekeeper. When it was meal time, Charley would provide Lemon with hot water and a bar of Palmolive soap and admonish him not to leave any black marks on the towel. He once instructed Mrs. Lemon on how to iron a shirt.

Bert Stillman homesteaded the land now occupied by Buck's T - 4, one of the few homesteaders in the canyon without a good spring. Stillman used to say that August was the best time of year to hunt for elk, and that they were better in the park than anywhere else. While Stillman may have bent the rules a bit for himself, he didn't bend them for his friend Big Lew Bart. Big Lew had been squatting on a piece of property near Twin Cabin Creek when he answered the call of his country and went off

Ira and Myrtle Verwolf. —U.S.D.A. Forest Service

to fight in the Spanish-American War. Big Lew thought that the years he spent in the Army would count toward his homesteading requirement. Stillman thought this broke the law, and he successfully contested Lew's right to his land. Stillman also successfully contested the right of Pearl Lockhart to homestead below the West Fork, claiming her property had minerals. In each instance Stillman lacked personal interest in the property.

To the north of Stillman, Len Verwolf, a member of a large family which emigrated from Holland, took up a homestead claim. Verwolf and his brother Guy took in many hunters during the season and served as guides. Another brother, Ira, worked as a wrangler at several of the local ranches, including some years at the 320 Ranch.

Two latecomers to the canyon homesteaded up Beaver Creek. Both Arthur Sherb and John Hinckley came into the canyon in the early teens. Sherb proved up in 1918. The Sherb family lived on their pleasant homestead in a twenty-one-foot by eighteen-foot cabin. Over the years the Sherbs had at least eight children. Old timers remember little about the Sherbs and soon after they proved up they sold their homestead and moved from the canyon.

Hinckley proved up and then went off to fight in the First World War. After the war he spent little time on his property, which he called "Bean Hill." When he was in the canyon, he usually prospected with Pete Karst

John Hinckley's cabin on "Bean Hill". —U.S.D.A. Forest Service

or worked with the Lemons at the Rainbow Ranch. George Lemon remembers Hinckley as an accomplished jeweler. He also served as the midwife for several of the Sherb children. During the winter Hinckley liked to ride the rails as a hobo.

Over on Porcupine Creek, the Waters family moved in to buy Tom Mickleberry's squatter's rights and prove up. Like most people in similar positions Foley Waters left the canyon in the winter. He satisfied the

Foley Waters' place, 1915.

government requirements by living on his spread during the summer months, when he could grow a crop. Foley left an account of a young homesteader's experience:

> I had finished High School in Bozeman and was a Freshman in the college there when my father, Dr. J. M. Waters, got the idea that any Montanan who failed to get title to some of that good Montana land was missing a golden opportunity, and suggested to me that we form a sort of partnership and take up some land in the basin. . . . It is true that my experience was not impressive. I could ride or drive a horse, was a fairly proficient camper and fisherman, and under sufficient pressure could milk a cow, tend a lawn or garden or saw the innumerable cords that our wood burning cook stove seemed to require.

I started out with all flags flying on June 19, 1900, some two months before my nineteenth birthday. The second afternoon I reached Tom Michener's place and since the river was high Tom had me stay there that night. Next morning he showed me a ford that he said was not too bad. A little more and it would have been too bad. . . .

Since I was too young to file on the land as a homestead, Father filed a desert claim, which required certain improvements and that a certain number of acres be brought under irrigation. But we were not dubious about being able to carry out the requirement, for we planned to make the desert blossom as the rose and have a self supporting ranch out of it. But on that first morning it did not blossom worth a cent as, after driving through the endless acres of sagebrush I had set out to subdue, I stepped down from my elderly wagon before that deserted cabin, sun baked and forlorn, with no floor, no windows except two whiskey flasks set in the door and filled with the rubbish that accumulates when people leave and the mice and mountain rats take over. I felt a long way from home. But from here on this was home, so there was nothing to do but spend the rest of the day in making it as hospitable as possible.

Foley Waters recalls having few visitors, but settlers were beginning to move into his area. Billy Lytle took up a claim to the south of

Billy, Pearl and Willabelle Lytle. —Willabelle Lytle Tonn

Porcupine. In 1910 Billy married Pearl Lockhart, who had tried to get a homestead title to land below the West Fork. Although blind, Pearl cooked so well for hunters who stayed with the Lytles that few people knew of her handicap. The Lytles built a two-story log home, which still stands by the Gallatin River.

Billy Lytle ran cattle for himself and valley residents. He also raised horses and later sheep. He used a Mormon derrick to stack his hay. Among other jobs he worked for a time in a sawmill.

Billy Lytle and his brothers Clarence came from Wisconsin. Over the years they continued to have their hometown newspaper delivered, one subscription for each brother. Old-time residents remember that when the paper arrived, Billy would stop right in the middle of cutting hay to read the news from Wisconsin.

The Jim Roans lived near the Lytles. They settled in the canyon early on and Emma Roan and Mamie Michener enjoyed cooking and sewing together while the men cut wood or put up fences or buildings.

Sam and Josie Wilson took up their homestead at the Buffalo Horn Ranch. Sam had come into the canyon on a hunting trip in 1897 with Jim Trail and Axe Johnson. The beauty of the area so impressed the three young men that they vowed to return together and homestead. In the summer of 1899 Sam Wilson brought his bride, Josie, to Buffalo Horn Creek.

Sam came from Iowa. Josie had grown up in the Gallatin Valley and was well suited to homesteading. Sam and Josie had two children,

Josie Wilson with an elk she shot.
—Museum of the Rockies Photo Archives

Clinton and Helen Eloise. Josie Wilson used formal terms of address, and everyone in the canyon, except the Micheners, called her Mrs. Wilson. She shot game for meat and cheerfully served home cooked meals to the many bachelors who frequented her table—the single men came to the homestead once a week after the Eldredge Post Office moved from Taylor Fork to the Buffalo Horn Ranch in 1906. Mrs. Wilson gave the men odd jobs then served them dinner, and they spent the evening trading news and stories. After staying the night in the Wilson's bunkhouse, they returned to their lonely cabins. Some of the bachelors traveled up to twenty-five miles on cross country skis or snow shoes to get a home cooked-meal, some news, and their mail.

Families always invited the bachelors for Thanksgiving. The men usually stayed right through Christmas, figuring it made little sense to go home for a few weeks. They helped around the camp doing chores and contributed fresh meat to the household. The families called the men "star boarders," but not to their faces.

When the Wilsons took up their homestead, several cabins were already on the property. The Wilsons constructed a log home using pegs rather than nails, and because of the cold weather or the difficulty in getting glass up the canyon, they installed only one window. This fine example of log work is now part of the large dining room at the 320 Ranch.

Sam harnessed the energy in Buffalo Horn Creek to run a wood saw. He used it to cut the many cords of wood needed to keep the ranch warm in the winter. Sam tried many schemes to make money, but he usually fell short of the mark. He brought a herd of wild horses from Nevada and let them range freely up Taylor Fork. When fall approached, sixty-five head drifted to the top of Cinnamon Mountain and starved to death in a snow storm. The other horses were unbroken and of little value when Sam had to sell the remainder of the herd. Sam then bought twelve hundred angora goats, which he planned to raise for their wool and kids. He put them in a shed to keep warm, but one bitterly cold night they piled together so closely that five hundred of the goats smothered to death. That ended his goat raising venture.

Sam Wilson's father homesteaded an adjoining quarter section, and the two homesteads together totaled 320 acres, from which the ranch got its later name. The Wilsons always referred to their homestead as the Buffalo Horn Ranch.

Axel Johnson, one of the trio who had hunted in the canyon in 1897, homesteaded what is today the Elkhorn Ranch. Axe came from Stockholm, Sweden, and lived up to his name by being a fine builder of log cabins. An excellent hunter and trapper, he could travel with ease on cross-country skis.

Jim Trail homesteaded on Monument Creek, just below the present Black Butte Ranch, and proved up in 1905. Like Johnson he hunted and trapped for a living. One story about Trail holds that one spring Beaver Tom, the trapper made notorious by Andrew Garcia in his autobiography *Tough Trip Through Paradise*, spent time with Jim Trail and taught him how to trap beaver. The story goes that Beaver Tom was so dirty that Trail refused to let him prepare a meal. Jim Trail later sold his claim to the Hercules Gold Company, organized by Tom Michener. Close by Trail, Lester Pierstorff proved up on his claim about 1912. Pierstorff worked at odd jobs around the canyon.

Up Sage Creek the Victor Adams family took up a homestead. Sadly, they had a child drown in the creek. Mary Sales from the Gallatin Valley had started the homestead but left the canyon when she contracted tuberculosis. Mary and Josie Wilson were best friends from high school.

Between Johnson and Trail, Dr. Safely, from Livingston, took up a homestead at Snowflake Springs. The doctor had two reasons for taking

Mary Sales. —Mary Owens

up this land in the upper Gallatin: he planned to build a sanitarium for diabetics because he thought they would benefit from pure mountain air, home cooking, and water from the year-round spring. And should the sanitarium fail, the doctor thought he would make money selling his property. Safely was convinced that the Montana Power Co. would put in a hydroelectric plant at Snowflake Springs to provide power for the canyon. However, the power company never constructed the dam.

While helping construct the sanitarium, Dr. Safely's step-son, Harry Hill, fell from the roof and split his head open. The carpenters dispatched Bert Stillman on his little black mare, the fastest horse in the canyon, to telephone for help. The nearest phone was at West Fork, enclosed in a glass case that read, "Break only in case of fire." So law-abiding Bert galloped past West Fork and on to Squaw Creek. There he found a United States Forest Service telephone not enclosed in a glass case. Meanwhile, the builders at the site, hoping to stanch the flow of blood, stuffed the wound with ashes and hurried down the canyon with the injured man. The wound healed, the young man suffered no lasting effects, and ashes became the material of choice in the canyon to stop the flow of blood.

Ed Alderson homesteaded closer to the park line. His family owned the *Weekly Courier*, and Ed set type for the paper. In the canyon Alderson served as a hunter and guide. In his obituary of 19 May 1917, the *Weekly Courier* said:

> As a guide in Yellowstone National Park and surrounding country he outfitted and took care of many parties . . . making friends of government officials and others of national importance. His knowledge of directions was faultless and he would relate instances of being out in the woods with hunters and explorers who, when it became time to return to camp, would insist on taking the opposite direction from that which would take them to their destination. Always firm . . . good natured, he would simply say, "You may go in that direction if you choose, I am going to camp." They followed him as a matter of course, but exceedingly skeptical as to the correctness of his judgement, only to find that it was unerring.
>
> Those who in later years had the pleasure of being his guest at his home in the West Gallatin Basin saw him at his best. They might find him old maidish and particular, everything was kept in such perfect order, but they left with the memory of a visit the charm of which they will never forget.

Around 1914 Alderson sold his homestead to the Hercules Mining Company. The mining company returned the land to him when the company failed to complete its plans to dredge the Gallatin River for gold.

Eventually Bud Story bought the Alderson place, the Lester Pierstorff homestead, and then the Trail property. Story then traded the Trail acreage with the U.S. Forest Service to consolidate his piece of property known today as the Black Butte Ranch.

Big Lew Bart had cabins up and down the Gallatin. While living near the park line, across the river from Specimen Creek, he worried about a rock on the ridge above him. It seemed poised to roll down and smash through his cabin. So Lew went up the hill and worked at the rock until it rolled down the hill and smashed the cabin. He prospected up and down the canyon and usually built a cabin next to his latest mining claim.

Eventually, Lew bought 120 acres of the Michener homestead and settled down on Mud Creek, to the north of the Stillman homestead. Forty acres of the original Michener land had been excluded from the homestead portion and claimed under mining law. Lew bought the homestead acreage, and the Micheners retained the mining claim.

Big Lew had peculiar eating habits. He believed in homeopathic remedies, perhaps because of his training as a pharmacist. Lew would eat just one or two foods for several months at a time. Once it was oranges and peanuts one month and cucumbers and apples the next month. He used at least two dozen different bottles of herbs daily. In later life he ate herbs for dinner. A story holds that he also ate fried gophers. Lew read voraciously and was always ready to try another experiment.

While building his place on Mud Creek, Lew had to haul his building logs quite a ways. He borrowed Old Smudge, the Lemon's work horse, and skidded out one log. The horse sweated so much that Lew felt sorry for him, so Lew dragged out every other log and let Old Smudge have a rest.

After Lew completed this cabin he built one farther up Mud Creek and tried to pull it down the road. He put a cable around the cabin, placed logs under it, and pulled it with his car. This method failed, so he left the cabin where it was and built another one on the north end of his property near the Gallatin River. He was without a nearby spring, so he used the Michener spring for water. He carried it in buckets suspended from a wooden yoke around his neck.

The Gallatin River bordered Lew's land near the West Fork, and he worried that the river, which was eroding its bank, was washing away his property. So Lew undertook to dig a series of ditches that channeled the water to the far bank and eroded that side of the river. Buck Knight remembers in the early years of the depression digging trenches for Big Lew for ten cents an hour.

Ray Michener tells how as a recent college graduate he took advantage of Big Lew one winter:

> As winter was closing in here came Lew one day. He brought his old car in and wanted to leave it for the winter. We told him that would be fine—we could use it for trips to town. We were kidding but he believed us. So he rigged up a travois like the Indians used to use to move camp. It consisted of two poles up over his shoulders and bound together behind. On this device he placed the battery out of his car and was going to drag it up to wherever he wintered. He was living at the time somewhere up above the 320. We were ashamed and took him back up and got his car. We stored it without using it.

At one time Lew served as caretaker for the Bud Story place at Black Butte. The ranch had a barn with a steep roof. Lew decided to fly off the roof in his "ultra light" flying machine. He fashioned a pair of wings, attached them to his arms, and rode a bicycle off the roof. Several people came from Bozeman to watch the experiment, which Lew did not repeat.

Unlike many of the other early settlers Bart never grazed cattle. Instead he operated a sawmill, constructed buildings and carried on his prospecting.

Pete Karst located above Moose Creek. In 1902 Karst started hauling freight and loggers for the newly formed Cooper Tie Company. In his travels up and down the canyon he found a stopping spot halfway between Bozeman and the tie camp. Charley Oliver had bought the land from Lew Bart for $375. Karst went into a partnership with Oliver but bought him out within a year and set up a place where travelers could spend the night or get a meal. Karst went on to develop a transportation company and dude ranch and to operate an asbestos mine.

One of the dudes who stayed with Karst in the early years loved the canyon enough to homestead. Charles High, a wealthy businessman from Ohio, homesteaded at Deer Creek, just over the Markley bridge.

Farther down the canyon at Castle Rock, Harris John Foster homesteaded. When he died, the land passed to his son Harrison, who wrangled at the Elkhorn and married Laura Banning, a guest at the ranch.

In 1911 the Sedgewick Benham family settled at Sheep Rock, beyond the growing colony at Spanish Creek. The Benham family had arrived in the colonies early in the 1700s. The family left Boston and over the next two hundred years slowly worked its way west across the United States. At age fifty-two Sedgewick came to the canyon with his wife Annie and their grown children, Earl, Nellie, and Carl. The family had a spread of 320 acres located on both sides of the river. It included pasture lands, a garden, and fields for hay and wheat. Father and son constructed sturdy log buildings to keep their stock and equipment. The Benhams had a bridge over the Gallatin River. Each spring the Benhams had to call on the Foster brothers, who lived upstream, to help remove the tree trunks that lodged up against the bridge pilings and threatened to wash them out.

Sedgewick Benham died at Christmas time in 1915. His wife and son Earl carried on the ranch for a while. Earl then married his cousin Bessie Benham, and the two moved to Buck Creek, becoming a part of the canyon community. In 1920 Earl and Bessie received a patent on one hundred acres. Daughter Nellie and her husband, Bert Shepherd, settled on the homestead at Sheep Rock, where they continued to ranch. The original Benham Ranch is now Rockhaven, a conference center owned by the Presbyterian church. The small building close to the chapel is the only original Benham structure left.

Just below Hell Roaring the Jim Burrows had a ranch used by many travelers who wanted a meal or a place to sleep. The original homesteader was a Mr. Eddy. In addition to cooking for travelers, the Burrows sold hay to hunters who spent the fall in the canyon.

Before the construction of the road few families spent time in the canyon. Bachelor prospectors roamed through the area during the summer, while hunters and trappers came in the winter months. Most of these men constructed crude cabins to serve as shelter. With the coming of the road, families moved in and set up farming and small lumber operations. They built homes, barns, bunk houses, and corrals. Most of the settlers continued to prospect, but usually the search for gold took a back seat to providing a stable family life.

Sitting on baled hay at the Waters' place: Lewis, Helen, Ray, Tom Michener, with Mamie Michener, others unknown.

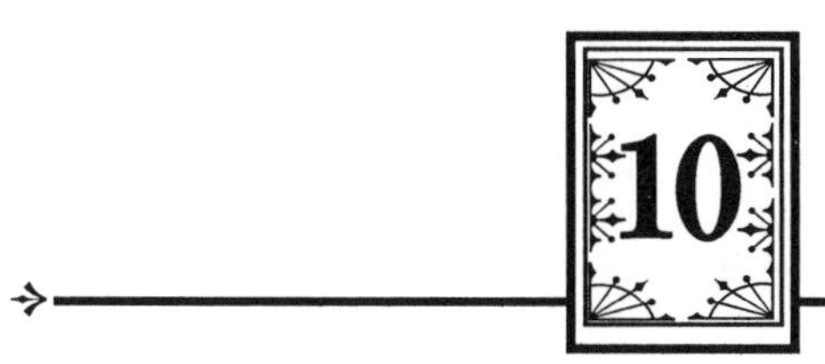

Everyday Life in the Canyon

Homesteading in the mountains of western Montana differed significantly from homesteading on the plains of the eastern part of the territory. Lured by gold, prospectors flocked to western Montana. They realized the economic potential of the area and stayed. The western homesteaders, almost without exception, had good water, plenty of wood for fuel and for building, and abundant game for food. In the mountains they grazed cattle and cut wild hay rather than attempting to farm. Rivers water many of the western valleys, the Gallatin Valley being a prime example, allowing valley farmers to irrigate their land and successfully raise crops. Homesteaders in the mountains took up land almost twenty years before settlement began in eastern Montana.

On the plains, settlement followed the railroad—the railroads encouraged homesteaders by publicizing the availability of inexpensive lands. The eastern homesteader practiced dry-land farming, depending on rainfall in an area historically prone to drought. The flat, treeless plains provided little water, wood, or game. Western Montana homesteaders reaped few riches but had an easier time than the homesteaders on the expansive plains to the east.

Soon after Montana became a territory, the government established a central survey point near Willow Creek in the Three Forks area from which to survey the entire territory. Mapping teams traveled the territory noting the rivers, streams, and mountains. Cartographers compiled master maps on which they drew neat squares, dividing the territory into townships of thirty-six square miles. Each township contained thirty-six sections measuring one square mile each and containing 640 acres. Each section was divided into four quarters, with each quarter section measuring one-half mile on each side and containing 160 acres. A quarter section formed the basic unit of homesteading.

On the prairie, government surveyors had an easy time putting surveyor marks in the ground for homesteaders to use when they claimed their land. In the mountains surveying became a more difficult job. On paper surveyors could still draw neat squares. In the field surveyors often found it impossible to run a straight line visually. Since the federal government did not complete the survey of the Gallatin Canyon until 1906, early settlers had to file preemption claims. The squatter claimed a certain piece of land containing 160 acres and marked it off by metes and bounds. This process required the claimant to show the corners in relation to prominent physical objects, such as large rocks, man-made piles of rocks, large fir trees, or clumps of trees such as aspens. Rivers and streams also served as reference points. Settlers had an idea where the section boundaries would be from consulting the map put out by the U.S. Geologic Survey. After the actual survey homesteaders had to adjust their boundaries to correct errors made in their metes and bounds surveys.

The federal government required homestead claimants to be U.S. citizens or to have filed citizenship papers. Single women could make

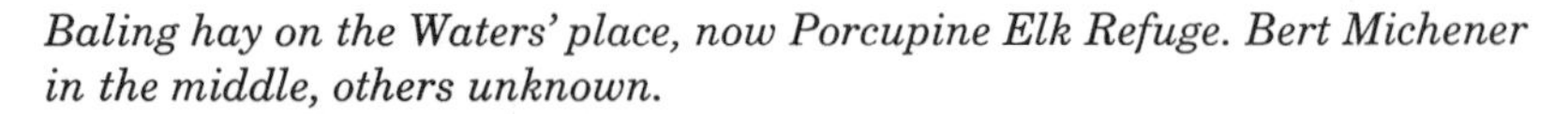

Baling hay on the Waters' place, now Porcupine Elk Refuge. Bert Michener in the middle, others unknown.

claims in their own names. Claimants had to pay a $10 filing fee and live on the land for five years, make improvements, and raise a crop. In the Gallatin Canyon settlers invariably raised wild hay, as little else will grow at the high elevation. Alternately, settlers could live on the land for six months and then gain title by paying $1.25 an acre.

Congress approved various versions of the homestead law. The Desert Land Act required the person filing to irrigate crops. Settlers on Porcupine Creek and Taylor Fork acquired land under this act. When the U. S. Government established the Gallatin National Forest it excluded lands good for agriculture and not needed for watershed. Many homesteaders in the Gallatin Canyon filed under the Forest Homestead Act of 1906.

Homesteaders in the canyon chose sites that had good water, grasslands for grazing, and areas where they could cut hay. Many settlers chose to locate around the West Fork of the Gallatin and in its beautiful drainage. Settlements along the upper Gallatin River stretched from Karst Camp to the Yellowstone Park line. Porcupine Creek and Taylor Fork each supported several homesteaders.

After their marriage in August 1898, Thomas Michener and Mary Elizabeth (Mamie) Lockhart moved to the Gallatin Canyon. The experiences of Tom and Mamie Michener typify those of mountain settlers in isolated canyons of western Montana. Before they left the canyon in 1919, the Micheners had six children: Maggie, Charley, Ora, Helen, Ray and Dorothy. Several of these children have left written and taped accounts of their early life in the Gallatin Canyon.

In September 1898 Mamie and Thomas Michener set out on the newly constructed road for their homestead. Behind their wagon they trailed their horses and a milk cow. They carried their winter supplies, a maple drop-leaf dining table, and a ten-volume set of the Ridpath World History in their wagon. The purchase of the world history started the large library the family collected over the years. The entire Michener family used and treasured the books, especially during the long winter months.

Mamie divided their one-room cabin into two rooms by hanging a curtain across a corner to make a bedroom. Dirt provided insulation for the tarpaper roof. Enterprising Mamie had brought along a good supply of oilcloth to tack to the ceiling to keep the dirt and dust from sifting through. The Micheners heated their cabin with a potbelly stove and with a fire in the little sheet iron cookstove called the "Sheepherder's Stove."

By 1900 the family had improved its home by adding a large kitchen, living room, and dining room to the original cabin. Tom built the additional rooms from lumber sawed at the mill brought in by his brother

Bert. The Michener home was the first non-log cabin in the canyon. The new white pine floors filled Mamie with pride. She would get a bucket of sand from the river and scrub the floors to keep them white. The men would then come home and track in dirt, and Mamie had to clean the floor again.

In summer, Tom's mother Emmaline and his sister Myrtle came to stay with the family in the canyon. Grandmothers and aunts helped in the warm months by making clothes for the children, doing the darning, and some cooking. In the fall the grandmothers, aunts, and uncles usually left, and the Micheners and other families in the canyon were alone, except for the bachelors who lived in the bunkhouse.

As the family increased and the workload became greater, Mamie hired help, especially in the summer and the fall. The family hired a cook to prepare meals for the many lumberjacks and hunters who stopped on their way up or down the canyon. Usually a couple would come together, she to cook, he to work with the men. Any children would play with the family children. Meals cost the travelers about twenty-five cents and provided cash to the family, who like other residents lived in a largely self-sufficient, barter community.

On Monday Mamie did the laundry. Before they left in the morning to do their chores, the men carried in water to fill the copper bottom oblong galvanized tub (called a boiler) and all the tea kettles, which they put on the stove to heat. An extra tub of cold rinse water sat on a bench. Homemade lye soap, though rough on the hands, cleaned the clothes well. Mamie first scrubbed the cotton clothes, towels, and sheets on a board, then boiled them. Next she rinsed the clothes in cold water, to which she had added Mrs. White's Ball Blueing. Mamie, like many settlers, took great pride in white towels, white sheets, and white diapers. In winter Mamie hung cotton items on the clothesline where they froze dry. In summer she laid them out on the grass, letting the sun whiten them as they dried.

The family wore winter clothing made of wool. Wool clothing is ruined by boiling and cannot be hung outdoors. Instead, homesteaders washed the items in warm water and dried them in a warm room on racks or the backs of chairs, steaming up the room and smelling as only wet wool can smell. Wash day in winter was not the best day of the week.

In summer the Michener women tried to get the washing done by noon so they could go fishing. Grandmother Emmaline and Mamie would take their cane poles and head up the Gallatin to a favorite fishing hole. Emmaline said it rested her to go fishing after washing. Ray Michener recalled his mother's prowess as an angler:

> She had a bamboo pole about twelve feet long—maybe ten feet—that she used together with some line, leader and hook. With this outfit,

> baited with grasshoppers, she was death on fish. She required a minimum of one, preferably three children to be along to catch grasshoppers and to gather up the fish as she threw them back over her head. The fish would come loose and travel some long distances. No fooling, fifty to one hundred feet behind her. Grandmother Emmaline was expert at this also.

On Tuesday the women ironed everything with irons made of solid iron, which they put to heat on the hottest part of the stove. Some had iron handles, and the women needed pot holders to use these. The better irons had removeable wooden handles. Mamie also baked bread on Tuesday to take advantage of the hot stove and made a New England boiled dinner or a pot roast of elk, which she cooked slowly all day.

The original cabin had a little "Sheepherder's Cookstove." It heated fast and made good biscuits, but homesteaders found it hard to regulate the temperature for making bread or cake. By their third year in the canyon, Mamie got a large Monarch Kitchen Range. It had a warming oven and two ornate iron shelves just above the cooking area. A water

Charley and Maggie Michener feeding chickens.

reservoir kept water warm for washing dishes. Atop the stove Mamie could make plum pudding and Boston brown bread to go with baked beans.

A spring house attached to the kitchen had a three-foot by four-foot hole cut in the floor, under which water flowed year-round. The water retained a temperature of thirty-four degrees and kept vegetables and milk fresh for days. The water ran out from under the building and through the yard, creating a small creek where the children loved to play.

A shed attached to the spring house held huge blocks of ice covered with sawdust, keeping the ice frozen for up to two years. The family used chunks of ice in the summer for ice cream or lemonade, which they usually served on Sunday afternoons. They had another ice house located next to the Gallatin River.

Mamie cooked or planned all the meals in camp. She made butter year round and in summer, when the cows gave more milk, Mamie made hard brick cheese using rennet tablets. She put milk in large galvanized tubs, keeping it at just the right temperature until it made curds. Mamie compacted the curds into round molds with a mechanical press, covered the cheeses with cheese cloth, and then put them on shelves to cure. The Michener family was proud of Mamie's cheese.

Like other settlers the Micheners raised turkeys, chickens, and a few ducks. The growing season is short in the canyon, so the garden featured lettuce, radishes, onions, cabbage, carrots, and rhubarb. Mushrooms gathered in the woods and dried on the south side of the cabin roof varied gravies and sauces and added a touch of elegance.

Bert Michener planted watercress along the Gallatin River, and homesteaders picked it year-round for salad. Like Johnny Appleseed with his apples, Bert Michener took slips of watercress with him wherever he went and planted them in streams. You can still find watercress growing in the canyon along the Gallatin and in many of its side tributaries.

One year Mamie and Mrs. Roan, who lived two miles to the south of West Fork, made a large batch of mincemeat. They made some pies and put them outside to freeze, then brought one pie in and baked it. This proved so successful that they froze all of their pies in winter so they would always be handy for company.

The Michener family filled the root cellar with vegetables from the valley: potatoes, carrots, onions, turnips, heads of cabbage, boxes of apples, and canned goods, both home-canned and store-bought. The homemaker combined apple juice with mother of vinegar to make a fine product. The family bought cabbage, cucumbers, and crab apples in the valley, putting them up as crocks of sauerkraut and pickles.

People without root cellars let their vegetables freeze and then used them frozen over the winter. This worked until the weather became warm, when they had to throw out the thawed vegetables. Dried food, however, lasted longer than fresh-frozen food.

Mamie made chokecherry syrup and apple jelly. After the first frost she picked Oregon grape berries and made them into jelly. Each fall the family made a trip into the valley to pick buffalo berries, which grow on a bush with silver-colored leaves and large thorns. The Micheners disliked picking these tasty berries, as they were hard to find, but they enjoyed eating the jelly with meat and fowl. Mamie Michener thought Thanksgiving dinner was incomplete without buffalo berry jelly.

The settlers prized huckleberries and chokecherries, and people refused to tell where they got their huckleberries. When gathering huckleberries one person had to keep watch for a bear in the patch while the others picked berries.

Mrs. Wilson of the Buffalo Horn Ranch and Mamie Michener got together in the fall to shoot and preserve grouse. They went to Cow Flats behind the Wilson's ranch and shot grouse, which they dressed, roasted, and placed in large crocks, packing the birds tightly together. They then poured melted lard over the grouse, making sure to remove all air bubbles. They topped off the crock with more lard and kept the crocks of grouse for several months in the spring house. They also preserved sausage this way, so they had meat year-round. When warm weather came, ham and bacon were the only meats that would keep.

Mamie learned to use bear grease to make biscuits and pie crust. The Michener's never ate the bear meat but rendered the fat like pork fat and used the lard in cooking. Homesteaders hung bear fat in a warm place, allowing grease to drip out. This grease remained liquid and homesteaders used it on leather boots and shoes and on harnesses to make them waterproof and soft. When men dressed up and went to town to sell bear hides, they slicked back their hair with bear grease.

At Sunday dinner Mamie Michener always used a white linen tablecloth and white linen napkins. Wildflowers decorated the table from early spring to late fall. Usually on Sunday afternoons in summer the family made ice cream, a real hot weather treat. Everyone had a chance to help turn the handle of the freezer.

Many home remedies contained turpentine. Early settlers used turpentine to dose horses, dogs, and people, pouring turpentine directly into wounds. For chest congestion homesteaders made a poultice of lard and turpentine, applied it to a length of cotton flannel, and affixed the poultice around the neck with a large safety pin. The ingredients of the poultices varied, with some women using lard with dry mustard and others using fried onions, but every remedy depended on flannel and a safety pin. Emmaline Michener made a poultice of one tablespoon of cayenne pepper boiled for ten minutes in one cup of cider vinegar, which must have wreaked havoc on the patient.

Homesteaders relied on a drop of kerosene on the tongue of a congested baby to bring up its phlegm. A salve for infections consisted of bar soap and sugar. Onion juice, sugar, vinegar, and butter made a cough syrup. A good tonic consisted of one pint whiskey, twelve tablespoons of olive oil, and eight tablespoons of honey. Stores carried horehound bark candy, which the homemaker made into a tea to thin out winter blood. The Micheners liked the taste and drank horehound tea year-round.

Homesteaders learned to deal with accidents as well as with illnesses. Ray Michener tells of losing his thumb:

> Helen and I undertook to do some carpentry work one day. I held a piece of wood on the chopping block, and Helen took the big double

Myrtle Michener with fawn.
—Museum of the Rockies Photo Archives

Bert and Tom Michener hauling firewood.

> bitted axe to it. She got about one-half inch of my right thumb. It was just hanging by the skin. We ran screaming for Mother. She grabbed the thumb, put it back where it belonged and bandaged it solid. It never did cause me any problem. The scar is quite visible to this day.

Margaret Michener was the first white child born in the canyon. In early July 1899 her father left "Old Fannie" the Michener horse at the Burrows' place above Spanish Creek for Dr. Waters to ride up the canyon. Though he arrived too late to assist in Margaret's birth, Dr. Waters stayed for three weeks and assured her a good start in life. For this long house visit Dr. Waters received $25. The doctor stayed in the canyon until the birth of Gladys Robol, whose parents lived just a short way up the road from the Micheners. Dr. Waters rode, and fished, and spent some of his time looking at property. In 1900 he bought the squatter's rights to the Mickleberry place on Porcupine.

After giving birth to Margaret in the canyon, Mamie Michener went to Bozeman for the birth of her other children, except for Ora, who arrived before her mother could get to the hospital. Ora likes to recount her debut:

> Then when I came along I was supposed to be born in Bozeman but things didn't work out the way they were supposed to and so I was born in the canyon, and my father went to get Mrs. Wilson. When Mrs. Wilson arrived soon after my birth, she found I needed to be warmed up, so she said, "I let the oven door down and put her on the oven door and went to take care of her mother, and when I came back the poor little thing was half baked."

A great closeness grew up between the few women in the canyon. Though they rarely saw each other, they took advantage of every opportunity to be together to cook or sew or do other household chores that went faster with two or three women working together.

The men on the ranch stayed as busy as the women but usually worked outdoors. Many ranch jobs changed with the seasons, but other chores stayed constant through the year. Morning and evening, summer or winter, settlers had to tend to the stock.

In winter Tom Michener rose early and dressed in 100 percent woolen clothing from underwear to shirts, pants, and sweaters. Over all these clothes he put a coat of canvas with a blanket lining. Summer and winter over his other clothing he wore a pair of bib overalls, made of durable sail cloth and marketed throughout the West by the Levi Strauss Company. Levis stopped the wind from blowing through woolen clothing and kept the ranchers' clothes clean.

Once dressed, Michener built a fire in the stove, lit his kerosene lantern, and went to the barn to milk and feed the cows and turn them out for the day. He fed the horses and turned out all but those he would need for work. Tom next fed the chickens and other fowl. In winter Tom opened the water hole, which had frozen over during the night, for the stock. Finally, he took the milk into the house, strained it, let it sit for an hour or two, and then skimmed off the cream. He was then ready for a hearty breakfast.

In the evening Tom once again milked the cows and put them, along with the horses and chickens, away for the night. In late winter and early spring ranchers always had to be ready to help in the birthing of calves and foals.

Homesteaders usually split firewood in the winter. The wood pile in the house needed daily replenishing.

A choreman, a valuable helper on the ranch, relieved settlers of some of the everyday work. He milked cows, fed chickens, cleaned barns, helped with the garden, and did some repair work. When a family went to town the choreman stayed at the ranch.

During the warm months the rancher's work changed, but the load never lightened. Milk cows still needed milking two times a day and the chickens still needed feeding. The rancher's only real relief was that he did not have to chop open the water hole.

The ranchers had to break horses. Ray Michener remembers the job falling to his older brother Charley, who also captured elk calves for sale to zoos:

> Charley was very adept at running down and catching elk calves. He would just jump off his horse and bulldog them. Charley was also our head (only) bronc tamer. He would ride horses out of that old corral

> down there on the flat. Sometimes he would have them snubbed up to a tame horse, sometimes not. That place down there used to be pretty swampy. He would ride them into the swamp and they were practically helpless. Or sometimes he would just let them go. Maggie said he was the best horseman in the world.

In winter Tom Michener sometimes cut and skidded building logs to the camp, often working with a hired man or with the neighbors. Tom's brother Bert brought a sawmill into the canyon, and the Micheners cut much of the lumber for the neighboring settlers. The few families in the canyon got together in the winter both for company and because two or more men together could do much more work than a man on his own. The men hunted, cut wood, and put up cabins while the women sewed, cooked, and watched over the children. The Wilsons, who homesteaded at Buffalo Horn, had to travel thirteen miles to the Micheners, so when they visited they stayed for several days.

The blacksmith shop operated winter and summer. The men made horseshoes and hinges, sharpened tools, and kept buggies and sleds operable. In the evening the men sat around the Michener's living room stove and washed down leather tack with saddle soap. They sharpened and replaced the handles of tools and made new gunstocks. Several men would work on elk horns in the evening, carving gun racks, clothes hangers, door handles, and cribbage boards.

The homesteaders also had to break trails through the snow to the neighboring ranches or to Karst Camp, so Pete Karst could deliver the mail. Even after the automobile came into use, canyon residents used horse-drawn sleds in winter for transportation. Autos had to wait to get out until the snow melted, or drivers traveled early in the morning so the car would stay on top of the hard snow and not fall through and get stuck.

With spring came the job of repairing the bridges damaged by the ice jams of winter. The south end of the Gallatin River freezes from the West Fork south for twenty-five miles. North of the West Fork it often freezes on the bottom with water running over the ice. February usually brings a warm spell with Chinook winds. These winds melt the snow and cause water pressure to build up behind the ice. When the ice breaks up, water pushes the two-foot to ten-foot blocks into great piles. The ice jams sail down the river, moving rocks and cutting trees from the banks. Like a rifle shot, the sound of popping ice travels for miles.

Road crews built the first bridges in the canyon low to the water, from bank to bank. One or more piers, built of logs stacked in a triangular shape and filled with rocks, created braces for the low bridges. Ice jams washed out one or another bridge every year. High water also piled branches and logs against piers and washed out bridges. Only at the end of June could homesteaders stop worrying about the damage that the Gallatin River might do to their bridges.

Trapped bear.
—U.S.D.A. ForestService

Every homestead family faced the problem of bringing in cash. To supplement his income, Tom Michener, like many of the other men in the canyon, often ran trap lines in late winter and early spring for marten, beaver, mink, weasel, and lynx. The trapper found beaver and mink close to home by the river. Marten, trapped high in the mountains and reached on skis or snowshoes, required great stamina but rewarded the trapper with high prices.

Michener checked his trap lines at least two times a week. He usually put a sandwich in his coat pocket and loaded up on extra traps and bait for a long, lonely day. He started at daylight and reached home after dark. Bert Stillman had cabins in the mountains one day's travel distance from each other. He spent a week at a time out on the line.

In spring homesteaders trapped bear. The men constructed cages out of poles, then baited and set traps in the cages. The huge iron traps weighed about forty pounds each, and it took two men to set them. Without the cage a bear would run away with the trap. Settlers used bear hides for rugs in cabins and a market always existed for the extra hides.

All year settlers hunted game for the table. In the early days men hunted for market, but by the turn of the century overhunting had made game scarce, and canyon residents hunted for their own use. Poaching went on in the canyon, mainly to supply meat for the table.

Summer found the ranchers busy mending fences, breaking horses, digging irrigation ditches, herding cattle, cutting hay, and doing a million other chores. Foley Waters and his father, Dr. Waters, bought the homestead claim of Tom Mickleberry in 1900. Foley gave an account of building fences and digging irrigation ditches along Porcupine Creek. Foley said:

> Tom Mickleberry had built a leaning fence with a pole on top and two barbed wires, using for posts a lot of abandoned railroad ties that lay along the creek. He fenced in a generous amount of territory but was rather niggardly with the number of braces on the fence, so some of my vivid memories of those days are of waking in the morning to find the place overrun with cattle from the range and of dropping my plans for the day to spend hours on the sagebrush-covered hills wrestling those heavy ties back onto their feet, restapling the wire, and putting in extra braces. As long as I was there I never got that problem fully licked.
>
> With proper ambition I wanted a ditch that would cover every available bit of land, so I started at the highest point of what seemed good plow land and worked back to the creek, a two or three days' job working alone, but further consideration convinced me that what I had laid out was a job for the Reclamation Service instead of for one medium-sized youth, so I lowered my sights and surveyed another line which resulted in the present ditch just above the cabin. Most of the dirt was moved with horse sweat and a plow but sizeable stretches called for man sweat and a pick. My back still aches when I think of those stretches.

Many people raised horses and cattle for valley ranchers, which meant they herded the cattle in the summer. Some herded cattle for large outfits such as Cunningham and Behring. The Micheners herded for John Wylie and his neighbors in the valley.

Foley Waters ran cattle for himself and others for several years. He left the following description of the independent cattle operations:

> Tommy Dolan and Enoch Sales were running Still Huling's cattle on Tepee Creek, John Kirby ran a mixed bunch up Taylor's Fork, Jack and Charley Smith did the same on Beaver Creek and the West Fork, and I with Frank Walker and Charley Ogle, then Len Verwolf, and finally Billy Lytle, ran another mixed bunch on Porcupine. We collected them in the spring, about May 20th, from owners in the valley, who had no summer pasture, and delivered them back in the fall after threshing, for which service we collected $1.00 per head for those returned, plus the cost of salt.

The ranchers cut wild hay and put it up in stacks for use over the winter. Tom Michener sowed timothy on irrigated lands along the Gallatin River and Porcupine Creek. Family members and a hired man or two did most of the hay cutting. Maggie and Ora Michener both remember driving the team for the stacker when the family cut timothy in Porcupine. Michener tried raising oats at his place on the West Fork, but they froze in mid-August.

In 1906 the Michener family began catering to dudes as a source of cash. Taking on dudes required a family member to saddle horses every morning for the day's ride. Usually the man of the homestead led the dudes on their excursions, taking them fishing and on picnics, and for supper rides in a hay wagon. Occasionally they would go for pack trips. This meant that the husband, and perhaps a son or daughter, had to load all the pack animals, put up and take down the tents, cook the meals, and saddle the horses for the guests. While the husband led the dudes around the countryside, the wife oversaw the cooking of the meals to feed the guests.

During the summer many men went into the valley to work to earn cash. Frank Blanchard ran big farm equipment, such as separators and threshers, for valley farmers. Bert Stillman irrigated in the valley, and Tom Michener worked as a surveyor laying out irrigation ditches.

In the fall, families in the canyon took in hunters, giving them room and board and sometimes acting as guides. All through the year canyon women cooked for freighters, loggers, herders, and surveying and mapping crews.

The prospect of hitting it big in gold kept many a family going through the early years. By the time the settlers realized there was no money in prospecting, they had come to love the canyon so much that they stayed on, knowing no pot of gold awaited at the end of the rainbow, and only hard work would sustain a family in the canyon.

Cowboys and Tie Hackers

The road constructed in 1898 allowed Bozeman entrepreneurs to move into the canyon to take advantage of the unregulated land rich in both grazing and timber. Large operations in both of these industries concentrated in the Taylor Fork area.

Marshall Seymour Cunningham and Hans Behring ran the largest and longest operating cattle outfit in the Gallatin Canyon from 1898 to 1923. Canyon residents ran smaller herds, usually composed of their own cattle and those belonging to Gallatin Valley ranchers, as a way of making cash.

Marshall Seymour Cunningham came from a prominent Virginia family. His grandfather served as an officer in the Mexican-American War, and his father served as a major in the Confederate Army. Cunningham, born in 1878, spent one year at the University of Virginia before heading for Montana. Gold fever soon overtook him and he traveled to Alaska, where he was the first white person over the Chilcoot Pass and where he got his nickname, "Klondike" Cunningham. He spent a short time constructing boats and running the rapids of the Yukon River before returning to Bozeman and going into the cattle business with Hans Behring.

Behring came from a family of bankers in Denmark. He reached Bozeman at the age of nineteen and took up a homestead (illegal at the age of nineteen), but he spent little time on his land. For four years Behring worked as a surveyor for the U.S. Topographical Survey. In 1898 Behring sold his homestead to Judge Armstrong for $2,500 and invested with Seymour Cunningham in 1,260 head of mostly Hereford cattle. Headquartered at Taylor Fork the new partnership of Cunningham

The Nine Quarter Circle Ranch in 1907. —Kim Kelsey

and Behring ran its herds into the Gallatin Basin before the federal government created the forest reserves and when grazing was free to the first person to get to it.

To establish the home ranch, Cunningham and Behring bought the Clyde Tedrick homestead for $2,500. They also acquired the Kirby and Marble homesteads. John Kirby had 160 acres from a regular homestead claim and an additional 160 acres of desert claim, for which he received title by irrigating land and grazing cattle on it.

Over the years Cunningham and Behring obtained their brands from different sources. In 1898 Cunningham and Behring bought between 800 and 900 head of cattle from Hiram Nixon, and with this herd they got the Nine Quarter Circle brand. In 1908 they bought the Rising Sun brand from Bill Ennis, founder of the town of Ennis, without cattle changing hands. They acquired the H Lazy P brand in 1910 from M. W. Penwell, along with five hundred head of cattle.

In 1900 Cunningham and Behring expanded their operations and took on nine partners, including valley cattlemen Ed Blackwood, Henry

Monforton, O. L. Ward, O. P. Davis, a man named Potter, and four others. This partnership lasted until 1903, when Cunningham and Behring bought out the Bozeman Valley cattlemen and ran their own operations until 1920.

Cunningham and Behring expanded its operation beyond the Gallatin Canyon. The partnership ran cattle in the Madison and Ruby valleys and on the Crow Reservation around Hardin. Besides grazing cattle in the Gallatin Canyon on Taylor Fork, the outfit also ran its cattle at West Fork, Tepee, and Daly creeks.

The Taylor Fork drainage proved large enough to hold both the Cunningham and Behring cattle operation and a tie-cutting enterprise owned by Walter Cooper. In 1902 Walter Cooper, backed by H. B. Holter and A. C. Johnson of Helena and Senator W. A. Clark of Butte, obtained a contract to supply ties and other lumber to the Northern Pacific Railroad. After completing the transcontinental rail lines in the 1880s, the railroads continued to construct feeder lines to serve burgeoning cattle and mining operations.

Cooper came to Montana from his home in New York State via the goldfields of Colorado. A prominent member of the Bozeman community, Cooper, over his long career was involved in mining, flour mills, rifle manufacturing, the fur trade, and water works in the Gallatin Valley.

Skidding out logs with team.

The Cooper operations began on Dodge Creek, named for hunter, trapper, and guide Ira Dodge. Cooper soon changed this name to Taylor Creek to honor his trusted foreman, J. L. Taylor. Taylor took care of all the bookkeeping, ran the commissary, and served as postmaster for the lumber camp.

Most of the Cooper logging operations took place on land bought from the Northern Pacific Railroad, with three tie-cutting camps up Taylor Fork. Subcontractors had one camp on Wapiti Creek in the Taylor Fork drainage, and several camps in the Buck Creek drainage farther down

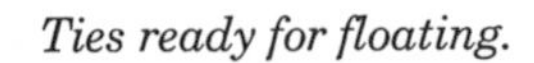

Ties ready for floating.

the canyon. Each of the main camps had fifty tie hackers, while the subcontractors had about forty men per camp.

Margaret Michener Kelly recalls that after the ground froze in the fall and the boggy places in the crude trail became less bothersome, but before the deep snows of winter arrived, an almost constant stream of freight wagons moved supplies up the canyon to the camps. Four to six horses drew each freight wagon. The loud rumbling of the wagon wheels on the frozen ground went on day and night. The wagons used the original 1898 road with its many difficult spots, the most hazardous being Sage Brush Point. Loaded wagons, team and all, often slid off the narrow, slippery road and rolled down to the river far below.

Loggers worked from this river boat to dislodge log jams.

Cooper constructed cook buildings and bunkhouses for the men. A commissary stocked lumber men's clothing and a large supply of snuff. The camps banned alcohol, and the cooks had a hard time keeping the men out of the vanilla and lemon extract. Full-time hunters provided meat for the workers. Cooper established a post office, named Eldredge for the postmaster general. When logging ceased the post office moved to the Wilson's Buffalo Horn Ranch where it remained until 1939.

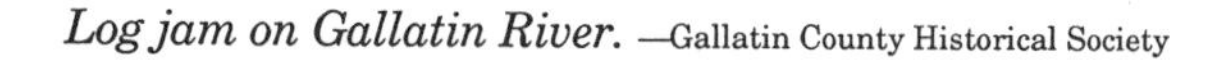

Log jam on Gallatin River. —Gallatin County Historical Society

Tie cutting proceeded all year with the men selectively cutting trees of the right size for railroad ties. The *Bozeman Weekly Courier* in 1924 estimated that loggers removed two and one-half million railroad ties and large quantities of other timber products during the five years that the Cooper enterprise operated.

Before the spring high waters the loggers dammed the creeks in the drainages where they cut ties. The lumberjacks piled the logs next to the creeks during the winter, and when high water came the loggers pushed the ties into the water, broke the dams, and watched the logs float away. Mr. Cooper's daughter, Miriam Cooper Bunker, gives the following account of the spring log float in a letter to Vic Benson in 1939:

> During the spring and summer of those years the railroad ties were floated down the Gallatin River to Central Park where they were shipped. Expert river men were hired for that purpose; most of them came from Canada or Oregon. The way they were able to handle the ties in that swift water was amazing and nothing short of miraculous. My father had two long river boats built which followed the ties and with long steel hooks the men in the boats untangled the tie-jams that formed. . . . The worst jam that I saw was at Cave Creek in those rapids there where the river is extremely swift and narrow and the rocks are very large. There were literally hundreds of ties wedged in around those huge rocks . . . in every conceivable shape and position. . . .
>
> The head river man was a man with only one hand. He had a hook for a hand and carried his long steel hook in his other hand. He was a wonder to watch. He leaned out of his boat and used both arms equally well, and could handle the boat better than any of the other men. The men shot all the rapids standing up in the boats, and it was a thrilling sight I shall never forget. A man would jump out of the boat on to a bunch of lodged ties, disentangle them, push them on ahead, then leap from one tie to another then onto the rocks, and from rock to rock till he reached shore, then run along the bank and catch up with the boat further down the river.
>
> During these spring drives . . . the work had to progress rapidly. Freight cars were waiting to be loaded at Central Park and the men worked on the river from daylight till dark. My father was constantly driving up and down the river on the look-out for saloon men from Bozeman who were constantly trying to sell liquor to the river men and smuggle it into the tie camps. He found this a constant annoyance.

The *Bozeman Weekly Courier* in July 1905, reported:

> The greater part of the men who have been employed by the Walter Cooper Company as lumber drivers on the West Gallatin River returned to Bozeman last Thursday [the 19th]. . . . The drive has been successful. . . . All the ties, to the number of 350,000 to 400,000, were

> floated down. . . . The low water came suddenly, especially down towards Central Park where the river is largely used for irrigation purposes. . . . Something like sixty men landed in Bozeman on Thursday and were paid off. . . . The making [of ties] continues in the mountains without interruption through the summer and the camp at the head of the West Gallatin is still a busy place.

The Cooper tie operation ended in 1907. The railroads had completed construction of most of the lines they were to build in Montana. The decreased demand for ties, coupled with a national recession, forced Cooper to cease operations, leaving his workers and suppliers unpaid.

Labor troubles started during the summer of 1907, when a hundred unpaid tie hackers marched down the canyon and camped in the middle of the road at Spanish Creek. Cooper called the sheriff to quell the disturbance, and the men dispersed. Still, the Cooper family kept a guard in their yard for the remainder of the summer. Cooper brought in other laborers to complete work at the tie camps. In 1907 Dick Dean took over Cooper's operation and conducted business on a much smaller scale until 1912.

One casualty of Cooper's operations went on to become a living legend in the Gallatin Canyon. Pete Karst had invested heavily in freighting equipment to haul men and supplies between Bozeman and Taylor Fork. The demise of the Cooper operation left Karst with a large debt. Seizing the opportunity as was his habit, Karst quickly went into dude ranching and mining, and became a success in the canyon.

The Nine Quarter Circle herd up Cache Creek. —Museum of the Rockies Photo Archives

Cooper turned over the lands he owned in the Taylor Fork drainage to H. B. Holter, of Helena, as payment for goods supplied. Holter in turn leased the land to Cunningham and Behring. When Cunningham and Behring increased the number of head of cattle in the Taylor Fork area, they tried to lease Northern Pacific Railroad lands to provide additional grass. With most of the Holter land grazed by cattle, the United States Forest Service opposed Cunningham and Behring's attempt to lease railroad lands. The Forest Service felt that the cattle would overgraze the drainage, leaving little grass for the elk. Cunningham and Behring recruited the Montana representatives in Washington to lobby for increased grazing acreage, and the state's representatives exerted considerable pressure on the Forest Service to drop its opposition.

Charles Anceney, Jr., of the Flying D Ranch was a good customer of the Northern Pacific Railroad. He had his partner Harry Child let the railroad know that Cunningham and Behring, too, would provide the railroad with plenty of business. The Northern Pacific, anticipating increased freight, agreed to lease up to seventy thousand acres of Montana railroad land to the partnership. Lease rent for a section (640 acres) was normally $50. Cunningham and Behring had the railroad lands declared half grazing and half timber, and the railroad reduced the grazing fee to $25 per section. The partnership leased sections in the Madison Valley and the Gallatin Canyon.

Cunnigham and Behring had the advantage over the years of access to cheap money. Cattle ranchers normally set up a line of credit with the local bank in the fall, drawing on the money over the winter, spring, and

Seymour Cunningham tying a calf. —Museum of the Rockies Photo Archives

summer and paying back the loan in the fall when they sold their cattle. Hans Behring had access to money in Denmark at 2 percent interest while other cattlemen in the state borrowed money locally at 16 percent interest. Easy credit allowed Cunningham and Behring to take advantage of sales of large herds of stock. In 1908 they bought 1,250 head of Aberdeen Angus cattle from W. W. Wylie at $80 a head.

Cunningham and Behring spent large amounts of time and money improving their land for cattle. Much of the land in the Cache Creek drainage, which runs into Taylor Fork, is steeply sloped. The company had to erect fences and grade inclines to keep the cattle from falling down slopes into the stream, damming up the water, and drowning. Poisonous Larkspur (*delphinium occidentale*) also posed a problem in the early summer. The cattle would graze on the plant and die within a short time. The outfit erected drift fences to keep the cattle from roaming into areas where larkspur grew.

Foley Waters, who homesteaded at Porcupine and herded cattle for valley ranchers, told how cowboys dealt with the deadly larkspur:

> The higher range where the best feed grew was infested with some sort of weed (we thought it was larkspur) which would kill cattle until it was fairly well matured, usually about the first week in July. Up until then we thought we had to keep the cattle down on the lower range which was mostly sagebrush with some short grass which was always overgrazed. It was a discouraging proposition for the cows and naturally they did not stay there willingly.
>
> We who were newcomers adopted the methods of the old-timers, that is to put the salt licks down low and get out every morning with a good dog and, when we found any adventuresome cows up where the feed was good in the poison country, hightail them down out of there to the salt licks. This fast trip down from the danger zone was supposed to make the poison work faster, so that any affected cows would fall and so give us a chance to apply our remedies. We had two of these. If the cow was not bloated we would cut an artery at the root of the tail until we got a good flow of blood. Strange as it may seem, it usually worked and the cow would soon be all right. If there was bloating the cow got the same treatment and also a puncture with a jack knife in the center of the hollow space between the last rib and the point of the hip bone. This released the gas in the paunch without serious danger of harming the patient. Sometimes if the bloating was severe it was hard to tell exactly where the proper point for the puncture was, but we had to take a chance and I do not remember that we ever lost an animal from the surgery. We found that in some cases the opening would close before the gas had entirely escaped, or perhaps more formed after the initial outrush. Like a good surgeon I was anxious to improve my technique, so evolved the scheme of cutting a six inch section of a large hollow weed and inserting it in the opening. I have seen more than one steer

get to his feet and go lumbering after his mate with the wind whistling through the hollow weed. I do not know whether it eventually fell out or in but next day it would have disappeared and the steer would seem hale and hearty.

When Billy Lytle became my partner in the herding business he came to some conclusions for which I give him credit, and which when put into practice greatly reduced our losses from poison. He reasoned that since running the cattle down out of the poison areas seemed to hasten or increase the action of the poison and so give us a chance to save them by our surgery, perhaps if we moved them down slowly the poison would not act at all and so they would not need saving.

We had also noticed that often an animal would seem all right until it reached the salt ground and began to lick salt and then would fall to the ground. So he argued that plenty of salt, instead of being an antidote to poisoning as we had been told by the old timers, might act to make it worse, and that we should not put out salt until the danger from poison was over.

There was a considerable area of relatively safe ground with good feed between the known poison range and the sagebrush country where we tried to hold the cattle, but we were afraid to leave them in this area because it was too easy for them to work higher up. We rode from the ranch house every morning and usually got back there for dinner at noon, so thought we had to drive everything clear down to the sagebrush so that they would not work up too high before we got back the next morning. Billy suggested setting up a camp on the range where one of us would spend full time, taking alternate weeks, and making a ride each evening just before dark as well as early in the morning. Eventually we established a camp in a grove of quaking asps [aspens] near the Twin Cabins Pass which enabled us to extend our safe range considerably. When we did this and quit running the cattle and quit salting them until the worst danger was over, we found our losses were greatly lessened, but unfortunately we had not much time to test our improved methods, for we stopped running cattle a year or two later.

In the fall, the partnership of Cunningham and Behring trailed its herds to Bozeman, over the Trail Creek Pass into the Yellowstone Valley, and on to the Pryor Mountains south of Billings to spend the winter. In the spring, cowboys trailed the cattle back over the same route. This trip usually took a month and a half. Mrs. Cunningham cooked for the outfit as it moved the cattle from winter pasture to summer grazing lands. When the herds approached West Fork, Mattie Cunningham and Mamie Michener spent two to three days together cooking for the cowboys and catching up on the news. They looked forward to the time they shared in the spring and fall.

The winter of 1919-20 devastated cattle operations all over the state. Almost every cattle outfit in Montana lost most of its stock in the extreme weather. The unusually dry summer of 1919 left rivers running too low even for valley farmers to irrigate. Farmers put up little hay in any area of the state. Grass in the canyon grew thin and sparse, and the cattle went to winter pasture with no extra fat.

Deep snows blanketed the range in early winter. A Chinook wind in January of 1920 melted the snow, but the freeze that quickly followed the warm weather covered the ground with ice. Heavy snows fell on the ice. Cattle unable to rustle any feed from the ice-and snow-covered range went hungry. Ranchers had little hay to feed them. Great numbers of stock died.

The laws of supply and demand dictate that the few cattle left would be worth much more than normal. But the bad winter coincided with a downturn in the national economy, which weakened the demand for meat. During the summer of 1920 cattle sold for about fifteen dollars a head even though it had cost $45 a head to get these same animals through the winter.

Cunningham and Behring operated with a large loan from the Madison Valley Stockman's Bank in Ennis, and this bank went broke along with Cunningham and Behring. Forced to reorganize, the partnership sold half of its stock to the Ringling interests from the White Sulphur Springs area, and the Taylor Fork Cattle Company became the Southern Montana Cattle Company.

The association with the Ringling interest placed Cunningham and Behring in an interesting position. In the early days of the Taylor Fork Cattle Company, Cunningham and Behring had taken a strong stand against sheep in the Gallatin Canyon. They had gone so far as to have State Representative Charles Hartman visit Gifford Pinchot, head of the United States Forest Service. Hartman emphasized to Pinchot the detrimental effect sheep grazing would have on the scenery and the tourist trade in the canyon. In the early teens Hartman convinced the Forest Service to prohibit ranchers from using the road to push sheep from Bozeman to the basin. The ranchers had to truck the sheep up the canyon at considerable expense. Now the new partnership of Cunningham, Behring and the Ringlings found it necessary to raise sheep to recoup its losses from cattle.

Hans Behring maintained in an interview with Dr. Caroline McGill that he always opposed bringing sheep into the canyon because of the damage they do to the grass and the watershed. Yet grazing sheep made economic sense, because the price of wool had remained stable through the national recession. Many cattlemen in Montana in the early 1920s ran sheep until they could get back on their feet financially, then they quickly returned to raising cattle. The Southern Montana Cattle Company brought the sheep into the Taylor Fork drainage over Indian Creek.

Behring also emphasized to Dr. McGill that the company never overgrazed the land. He told Dr. McGill that he would buy more cattle, and Cunningham would then go out and get more grazing land. However, the battle between the cattle interests and those sportsmen who wanted to preserve grass for elk went on for years and became more heated when the partnership went bankrupt after the extreme winter of 1919-20 and decided to bring sheep into the canyon.

In 1923 Cunningham and Behring dissolved their partnership. Behring became a state land agent in Helena. Seymour Cunningham dabbled in the dude business at the home ranch on Taylor Fork until his death in the 1930s, when his son Robert took over the operation. The Cunningham family sold the ranch in 1945 to Howard Kelsey, whose family still runs a dude ranch under the Nine Quarter Circle brand.

As the Cunningham and Behring partnership went out of business, the Charley Sappington family, from the Three Forks area, moved into the canyon. In 1920 the Sappingtons leased several sections of land from the Northern Pacific Railroad and ranged a herd of cattle in the Beaver Creek drainage. In the 1940s the family bought grazing land from the railroad. Their nephew Faye Parker still runs about three hundred pair on leased land on Beaver Creek.

The Gallatin National Forest

The Louisiana Purchase of 1803 brought most of the West under United States ownership, but exploration and settlement came slowly in many regions. The area encompassed by Yellowstone Park was relatively unknown until the early 1870s. The Washburn Expedition of 1870 explored and mapped the region, as did the Hayden Expeditions in 1871 and 1872. The unique geologic formations and the many geysers so impressed the members of the Washburn Expedition that, with the backing of the other members, Nathaniel Langford set out on a national lecture tour exclaiming the wonders of the area. He suggested that the federal government put the entire area under its protection as the nation's first national park. In 1872 President U. S. Grant signed the bill creating the park.

The federal government owned many more millions of acres of land in the West, which it planned to turn over to the emerging states or to sell to entrepreneurs. Certain of these lands held valuable timber, grass, water, and game animals, and many were rich in recreation potential. The westward movement of the population started a debate on the disposition of these large tracts of land.

The debate culminated in the creation of the Forest Reserves Act of 1891. The act specified that the United States Government would keep ownership of large tracts of land, mostly in the West, rather than selling them or turning them over to the states. The public would profit from these lands through the sale of timber, the leasing of grazing rights, the protection of watershed, and the retention of recreational lands. However, the Forest Reserves Act covered only some of the land in the West.

Much developable land had already been homesteaded and incorporated under local governments, and the government gave the railroads many thousands of acres of land as an incentive to build rail lines across the continent. The government also deemed some land to have little value for timber, grazing, watershed, or recreation and excluded it from the federal reserves.

Congress failed to provide a mechanism to manage public lands under the Forest Reserve Act. The lands drifted along under the Department of the Interior until 1896, when President Grover Cleveland appointed the National Forest Commission. With a budget of $25,000 the seven commissioners, all scientists, spent three months touring the West. They designated lands in several western states, including Montana, for inclusion in the federal reserves. The committee also issued the *Report of the National Forestry Committee of the National Academy of Sciences Upon the Inauguration of a Forest Policy for the Forested Lands of the United States*. Such a long name indicates a weighty report and the committee gave the public just that. The report provided for management of public lands under the Forest Management Act.

Leaning heavily on the experience of foresters in Europe, the report called for establishing additional national forests and for the Forest Service to manage all such forests "under the philosophy of sustained use and multiple yield. . . . for the greatest good for the greatest number in the long run."

The Forest Reserves Act based forest management on the philosophy that timber is a crop to be harvested and replanted. The government would manage the forest for the good of all, rather than sell the land to private interests that would make huge profits on timber. The act charged the Department of the Interior with protecting watershed, preventing fire, prohibiting overgrazing, and barring timber cutting by settlers except for personal use. The act excluded from national forests land suited to agriculture. This separation resulted in the Forest Homestead Act, under which most settlers in the Gallatin Canyon homesteaded.

In 1905 Congress transferred control of the forest reserves to the Department of Agriculture and established the United States Forest Service. The federal government put the six areas composing the Gallatin National Forest under its protection from 1899 to 1907. Area Six, the final phase, encompasses the Gallatin Canyon. The federal government included the land encircling Yellowstone National Park in the reserve system to create a buffer zone to protect the park and provide additional forage for game animals.

At the time the government created the reserves, the Northern Pacific Railroad already owned almost every other section of land within fifty miles of the track. This situation occurs in only a few of the country's national forests, and it presented problems in the government's efforts to control grazing and preserve grasslands for game. During the first half of the twentieth century, the Forest Service made land trades with the Northern Pacific Railroad in areas surrounding Yellowstone National Park to ensure forage for the elk that migrate from the park into the canyon during the winter months.

Early in the debate on the disposition of the national lands, a split occurred between Gifford Pinchot, the most influential and well-known member of the National Forest Commission, and John Muir, who was not a member of the committee. The articulate Pinchot received his training in forestry in Europe, and he held to the idea that individuals should harvest and treat forests like any other renewable resource. Muir argued that wilderness should remain in its natural condition, protected from mining, logging, grazing, and other non-recreational use. Pinchot won the debate, and the Forest Service still follows his plan.

Gifford Pinchot realized the importance of bringing the remaining large tracts of land in the United States under the control of the U.S. Government before developers and businesses stripped them of their resources. When Teddy Roosevelt became President in 1901, the two men worked quickly, without the approval of Congress, to put the

Sheep grazing on U.S. Forest Service Land. —U.S.D.A. Forest Service

Colonel and Elers Koch camping. —Museum of the Rockies Photo Archives

remaining public lands into national forests. By 1905 Congress had inserted itself into the decision-making process, but by that time Pinchot and Roosevelt had incorporated into national forests most of the remaining forested land in the West.

Elers Koch, for many years an employee of the United States Forest Service, explains how Roosevelt and Pinchot accomplished their aims:

> It was probably the best and most interesting job there ever was in the Forest Service. A man was given a state map of California, or Montana, or Idaho, with a green-colored block indicating the general area he was to cover. The first job was to go to the local Land Office and take off on township plats the status of the land. At the same time rough copies were made of the drainage and topography from the township plats of such if the area was surveyed.
>
> Sometimes a USGS map was available, or some sort of a county map. Then the examiner proceeded to ride the country and see it for himself. The area was covered usually at the rate of about two days to a township, and we really saw and rode through every township. [A township is six miles by six miles, or a square containing thirty-six square miles.] A rough type map was made showing the general classification of the cover [timber, grazing, watershed, etc.]. If there was no map, the examiner made a map as he went along.

Each man worked alone on a separate unit. He was allowed complete latitude as to how he covered the job. He might engage a packer with pack horse and saddle horses, or he might ride the country on a saddle horse, stopping at ranches, sheep camps, mines, or whatever offered when night overtook him. It was surely enjoyable work, each man his own boss, and seeing new country every day. The mapping was expected to be only approximate and there was no great amount of burdensome detail to worry about.

Considering the rapidity with which the work was done, it is surprising how well the original work has stood up. Most of the boundaries so established had little modification in subsequent years.

We spent the summer and fall in the field, and in late fall all repaired to Washington with our notes and maps and spent the winter getting the data on paper. As fast as a unit was completed a proclamation was drawn and sent to President Roosevelt for approval. It was a quick and efficient job, and before Congress got around to repeal the authority of the President to proclaim Forest Reserves nearly all the

Rhesis Fransham, first forest ranger in the canyon. —R. Evans

> remaining forested public land in the West had been safely covered into the Reserves. . . . In 1904 I covered the Gallatin. . . . I do not recall much opposition which developed while the examinations were being made. I guess it was all done so fast that the local people didn't know what was happening till T. R. signed the proclamation.

Once the government placed the land in the national forest system, it was subject to regulation and required the forest ranger to prevent illegal timber cutting and to issue permits to cattle and sheepmen. Koch continues:

> It is interesting to speculate on the amount of riding, camping out, and spitting and whittling the early rangers must have done to get all of these established grazing users under permit for the first time. Resistance must have been high and many hostile words were probably exchanged during the process. Yet, all of these animals were placed under permit by 1907.

Well before the federal government established the Gallatin National Forest, stock grazed in the canyon. The Murray brothers ran horses on Buffalo Horn Creek in the 1860s. In 1880 Lewis Michener drove a herd of young horses over Jack Creek and down the West Fork to summer. W. W. Wylie wintered many horses at Spanish Creek. After 1907 local rangers followed Pinchot's guidelines, giving preference to local ranchers for grazing permits.

The Forest Service allowed homesteaders and miners to cut timber for personal use on national forest land. Forestry officials had to patrol continually against unregulated commercial use of the forests. The extensive logging which took place in the canyon in the early part of the twentieth century was generally carried out on private property.

From the start of the national forest to 1940, the demand for both timber and grazing in the Gallatin Forest steadily decreased. The demand for ties by the various railroad companies declined, and the economy went into a recession. Forest Service records show that from 1910 until the late 1940s demand for the so called "inferior lumber" from the lodgepole pine that grows so well in the canyon fell steadily. Records also show a steady decline in grazing permits in the Gallatin National Forest from 1909 to 1939. In 1909, 18,365 head of cattle and 2,352 horses grazed in the forest, along with 161,012 sheep and 500 goats. By 1939 the numbers had dropped to 7,656 cattle, 1,189 horses, 63,868 sheep, and 16 goats. (The Gallatin Canyon comprises a small part of the Gallatin National Forest, so not all of these animals grazed in the canyon.)

The decreasing demand for grazing permits and logging rights still left forest rangers with little spare time on their hands. As the public became aware of the wildlife endangered by overgrazing, its attention turned to the elk in Yellowstone National Park and in the national forests surrounding the park. Forest Service rangers added the job of game warden to their previous duties of regulating timber cutting and grazing.

Mr. and Mrs. Fransham with friends at the first ranger station at Cinnamon Creek. —R. Evans

Students and teacher in 1936 at the Ophir School, built by Eugene Crail in 1928.
—U.S.D.A. Forest Service

The Little Log Schoolhouse

Homesteaders organized the first school district in the canyon in 1886. The little schoolhouse sat in a pasture near Spanish Creek, and residents held school there on and off for several years. The county organized several other schools over the years in the Spanish Creek area. Tom and Mamie Michener sent their oldest daughter Margaret, the first child to reach school age in the Gallatin Basin, to live at the Wylie Ranch at Spanish Creek so she could attend school. As she remembers, she did not stay long. "I started to school at Spanish Creek when I was barely six years old and went to school for nearly three whole weeks before I got so homesick for Old John, the horse, that I had to come home," she recalls.

By 1908 five children of school age lived in the Gallatin Basin: Margaret and Charley Michener from West Fork and three Stevens children—nieces and nephews of Mr. Blanchard at Dudley Creek. Their parents began to push for a school. Through 1908 the Gallatin County Commissioners decreed that residents had to organize and hold school for two months before the county would form a school district or provide money to hire a teacher, putting the burden on parents to finance the school and causing dissention between homesteaders and outside cattlemen who owned or leased numerous acres of grazing land in the canyon. The cattlemen whose families lived outside the area resented the idea of paying either school taxes on land they owned or higher grazing rights on land they leased.

Despite opposition from the cattlemen, the parents and bachelors who lived in the canyon raised $20 to hire a teacher for a two-month summer school. Relatives in Bozeman donated used books, and Bert

Stillman made desks. Frank Blanchard made a long work table and benches for outside use. Pearl Lockhart, Mamie Michener's sister, offered the use of her one room homestead cabin, located a quarter mile below the mouth of the West Fork. The cabin, set among willows, lodgepole pines, and quaking aspen, was situated halfway between the Michener and Blanchard homesteads.

Miss Kathleen Cope who lived in Bozeman, came to the basin to teach that first school. Kate hastily embarked on what turned out to be a pleasant two months.

> About a week before the final play at Logan [a small town near Three Forks], I heard of the little summer school at basin—only five pupils and they were all small—and what pleased me most, it was right in the heart of the Rockies. When it came to this last it didn't take long to decide. I packed all my dressups in my trunk, and what old duds and leftovers I could find in my grip, and storing my trunk at Logan, I took my grip and the stage for basin. . . .
>
> Yes, it was a happy summer. Every day brought something new, and the time flew all too quickly, for it seemed such a short time until summer had completely passed away and it was time for me to go home.

The first school in the canyon with students: Charley Michener, Ruth Stevens, Maggie Michener, two Stevens children, and Kate Cope in front of Pearl Lockhart's homestead cabin.

I loved teaching these children. They seemed to be so willing to learn and so happy to have a school. Though my wages for the summer were only $20 it was one of the finest summers that I have ever spent and I want to go back.

Summer school about 1912, Charley and Maggie Michener, Kate Cope, Ora Michener, and Evans Kelly.

Margaret Michener recorded her memories of Kathleen Cope and that first school in the lower basin:

> I wonder just what she felt when she first saw her school. One tiny cabin, dirt floor, dirt roof, one window, all located, however, in a beautiful spot beneath the trees. Around the table made by Mr. Blanchard we had work and nature studies that I'm sure must have surpassed much of that taught in our modern schools with teachers especially trained for that work. There the beauty of God's handiwork was pointed out in a way that few people could suppose.
>
> What a wonderful teacher she was. The children loved the tiny dark school room because she was in it. Every morning when it was fine they sat around the table and sang lustily "Pretty buds and flowers in your bright blue dresses." When it was raining or too chilly they sang inside sitting at the homemade desks and seats made by Mr. Stillman.
>
> In July the whole basin went down for a picnic supper. Wagons and saddle horses were tied to the trees all around the little school house.

The work table became a picnic table and boxes and cloths spread on the ground made other space for seating. When it grew dark, coal oil lanterns were brought out. Much merriment went on in the form of foot races and other contests, and later in the evening huge tales were spun around the big bon fire. Everyone went home much satisfied with their little school.

Shortly after the residents of the canyon had done all the work to organize the two-month school, the U.S. Congress ordered that all land in the country be divided into districts. Had the canyon residents waited another year, they would have avoided the problems that developed between the cattlemen and homesteaders.

Despite authorization from the U.S. Congress, the basin delayed organizing another school until 1912. In the intervening years the Michener children went to Bozeman for the school year. Josie and Sam Wilson of the Buffalo Horn Ranch either taught their children at home or sent them to school in Bozeman. Because the Wilsons lived twelve miles from the main settlement in the canyon, their children never attended canyon schools.

In 1912 residents formed Ophir School District Seventy-two. Voters elected Thomas Michener, J. H. Deever and Y. J. Kyunders trustees. These men held their first meeting at the Michener Camp on 18 December 1912. The board voted to name the school district Ophir—the city of gold mentioned in the Bible—because they treasured the new school district as they treasured gold. The first teacher in the new district was Miss Agnes Foster, the daughter of Mrs. Burrows, who operated a stage stop at Logger Creek. Miss Foster received a salary of $50 per month.

During the early years the district held school at various ranches or wherever someone made available a cabin. The teacher always boarded with a family. The Micheners, with six children, often hosted the teacher and the school. The number of students during the years 1913 to 1940 ranged from two to ten.

If the canyon had two children of school age, then the district qualified for a teacher. But, even though the canyon had more than two school-aged children, the district could not hold school in the winter. The families were too spread out for the students to get to school, so the parents usually taught their children, and any of the neighbors' who could reach the homestead. George Lemon remembers going to school one winter at the Benhams at Buck Creek. Bessie Benham taught the children and George traveled to and from school in a horse-drawn sled that his father built to carry hay for the horse.

If the district failed to hold school for several years in a row, the county could close the district. One winter, to ensure school was held so the canyon could keep the district, the Micheners left Charley and Ora on the West Fork with the workmen, the cook, and the school teacher while they went to Bozeman for the birth of their last child, Dorothy.

In 1919 a flu epidemic closed schools in Bozeman, giving Miss Bertha Rich the opportunity to teach in the Gallatin Canyon. Bertha explains:

> Early in January 1919, Miss Daisy Forest, county supervisor of schools in Gallatin County, said to me, "You know, we have a little mountain school about 45 miles from here that needs a teacher. There are only 5 pupils at the present time but those folks pay taxes and are entitled to a school. I had thought about whom I might send and my mind always returns to you." I'm sure that she never could really know how delightful the whole thing sounded to me as she unfolded her plan and how glad I would be to exchange my large school of fifty pupils in the valley for this little mountain school up the canyon. Because of the flu epidemic which raged through the land all the schools in our district were closed. Many of the teachers, having taken an emergency medical course given by the American Red Cross, had volunteered their services as practical nurses in homes. I had spent the previous three months in this way.
>
> I knew nothing of this little mountain school firsthand. However, during the course of Miss Forest's conversation I learned that the school would be held in one of the cabins of the Michener camp. This camp was a dude ranch in the summer and they also did placer mining. This sounded intriguing. "If there are any heavy snows," she said, "you might be snowed in for several weeks at a time." My, how exciting! I'd read of such things and was anxious to experience it for myself. So one afternoon in the middle of January I found myself being introduced to Mrs. Michener and Charley, her son, by Mrs. Wylie, a relative of Mrs. Michener and a friend of mine with whom I had been staying. She was the only one who encouraged me to go.
>
> When we arrived at camp, Mr. Michener, Helen, Ray, Dorothy, and several others came out to the car to meet us. If I felt a bit strange at first, that feeling soon wore off as I was taken in with a warm welcome. No fuss was made, I was just accepted.
>
> At supper a warm and friendly atmosphere prevailed with a good deal of joking, laughing, and exchanging of experiences. I didn't enter into so very much, personally, I was just so busy trying to get things straightened out. I think I shall never forget how I decided I wanted to stay there. Perhaps it was when I saw little five-year-old Dorothy get down from her chair and go around and stand by her dad's knee. Although he was busy talking he looked at her and said, "Well Dots, what is it? Want your story now?" She nodded in the affirmative so he set her on his knee and told her a story making it up as he went along, using her for the heroine. These stories he told were practically all animal stories and Dot's blue eyes would grow bigger and bigger as she listened to the wonderful parts she always played in the tale.

> So, with my mind made up to gain the most from my experiences, there I was, ready to launch out upon what proved to be two of the happiest years of my life.

Before Bertha Rich arrived, sixteen-year-old Ora Michener taught during the flu epidemic to keep the children up to grade.

Each school year the county superintendent from Bozeman paid a visit to the Ophir School. Between 1912 and 1918 Miss Ida Davis visited the school in her capacity as superintendent, bringing along a health officer to check the children's eyes. Miss Davis distributed samples of Colgate toothpaste to the students and instructed them in the proper way to brush their teeth.

R. B. Lemon taught in 1922 at Porcupine. With a credential for teaching college students, he spent the year with primary-age children. The Michener children rode horseback to class until the weather got too bad, and then their mother schooled them at home. Dorothy Michener Vick remembers reading *The Five Little Peppers, Black Beauty,* and Dickens, and making flash cards for arithmetic.

In 1923, after receiving teacher's training at college, Ora Michener Lemon taught school at the Lemon's Half Way Inn. Ora taught three little girls: Wanda Lemon, Willabelle Lytle, and the ranger's daughter, Virginia Callahan. Willabelle walked from the family homestead above Porcupine Creek to the Lemon's Ranch, always arriving late. Every Friday the students took a nature walk to observe the signs of the changing seasons. They often drew and painted from nature.

Reugemer, Cobbler, Michener, and Johnson students in 1920.

Often canyon families hired a teacher for the summer to bring the children up to grade. In 1923 the Michener and De Paugh families shared the cost of a summer teacher. Micheners paid the teacher's salary and provided a schoolroom, while the De Paughs furnished room and board. When the Wilsey family lived at the Nine Quarter Circle Ranch during the 1930s, Joan Wilsey rode horseback to the Benson's Covered Wagon Ranch for classes. The Wilseys and Bensons shared the cost of the teacher for Joan and Vic Benson.

In 1921 canyon families began to search for a permanent home for the school. Because of the distances between homesteads and the difficulty of travel in bad weather, every family with school-age children wanted the new school located nearby. Canyon residents continued to wrangle over the site of the school for the next seven years. In 1928, with canyon families still disagreeing on a location, the district superintendent instructed the school board to choose a location or lose the district. The trustees compromised and selected a site alongside Beaver Creek on the east side of what is now U.S. Highway 191. Eugene Crail built the one-room log cabin that served as the school until 1963, when the Gallatin Canyon Women's Club, using money bequeathed to it by Dr. Caroline McGill, built a new school.

The first teacher to use the schoolhouse built in 1928 was Adelaide McMullen. She taught Rachael and Buck Knight, George Lemon, and Peggy and Jim McMullen. The McMullen family lived at West Fork and had to walk the rough road to Porcupine everyday.

Between 1920 and 1940 the core of the school board consisted of William, Pearl and Clarence Lytle, Frank Blanchard, Jack and Mabel Wood, and Mrs. E. H. Benham. Other residents served as trustees, but for short times.

As a trustee, Bessie Benham placed great importance on penmanship. In 1920 she moved that the clerk notify the county school superintendent of the qualifications demanded of any teacher for the local school. The Ophir trustees required that the teacher give lessons in the Palmer Method of Penmanship, completing ten lessons each month of the school term. In 1921 Mrs. Benham moved to withhold $10 from the teacher's monthly salary until each student had completed and passed the Palmer Method. The district has no record of ever docking a teacher's pay for failing this requirement.

As some readers may remember, the Palmer Method used the entire arm from shoulder to hand. The student scrolled countless circles, pointed waves and teardrop-shaped loops across a paper, taking care to stay between the solid lower line and the dotted middle line for lower-case letters and between the solid lines for upper-case letters. Small d's and t's ended in a no man's land between the dotted line and the solid line.

In addition to penmanship the children studied reading, grammar, literature, spelling, and math. Teachers gave instruction in drawing and painting but not music. Each holiday saw the production of a traditional play or pageant.

Community activities centered around the school. Canyon residents held many potluck dinners, dances, and community meetings there, and the government held elections at the school. Eugene Crail built the original one room school with two cloak rooms, one on each side of the door, which voters used as booths on election day. Later, settlers converted these closets to the boy's bathroom and the girl's bathroom. When election day came the voters cast their ballots in the former closets, now bathrooms. Gallatin Canyon may have had the only voting booths in the United States located in bathrooms.

The enrollment at the Ophir School remained small until the 1950s, when the Corcoran Pulp Company moved into the West Fork drainage. The number of students then approached eighty and school was held at the B Bar K Ranch. When the pulp operation left the canyon, school enrollment reverted to a few pupils each year. Enrollment finally approached eighty students again only in the 1990s.

Dude Ranching

Canyon residents had long hosted hunters in the fall as a way of bringing in much-needed cash. Bozemanites, known as "pot hunters," would either camp or stay with families in the canyon for as long as a month while they laid in a supply of meat for the winter. Area families also supplied housing and horses and served as guides to wealthy sportsmen from the East. As more people from large cities visited Yellowstone National Park, stories of the abundant game in the area spread among sportsmen. They began to organize hunting trips to areas such as the Gallatin Canyon. These hunters provided a good source of revenue, and the residents of the canyon decided to expand into the summer dude ranch business to increase their cash earnings.

W. W. Wylie, an uncle to Mamie Michener, owned the camping concession in Yellowstone Park. In 1905 Wylie received a letter from Mr. F. O. Butler of the Butler Paper Company in Chicago. Butler inquired about a western camp where his family and guests could spend the summer riding and fishing. Wylie sent the letter to his sister Belle Lockhart who ran the Wylie Ranch at Spanish Creek. Mrs. Lockhart, busy looking after the staff and horses at Spanish Creek and hosting overnight guests on their way to and from the park, thought of her daughter and son-in-law, Mamie and Tom Michener. The Micheners had several cabins that would suit dudes, plus cooking facilities to feed a crowd. Tom Michener had an outgoing personality and loved to show the canyon to everyone who came through. So Belle Lockhart sent Butler's letter to the Micheners. Thomas Michener saw great possibilities in the dude ranch business. It delighted him that people wanted to come to see the beauty of the canyon.

Dudes enjoying ice cream at Michener Camp, 1907: Mrs. F.O. Butler, Julius, Paul (sitting on ice cream freezer), unknown woman, Mrs. Jones, and F.O. Butler.

Michener contacted F. O. Butler, who came by train to inspect the facilities at Michener Camp. Butler traveled up the canyon from Bozeman by Karst Stage. He liked what he saw, so Mrs. Butler with their two boys, Paul and Julius, and another couple came to spend the summer of 1906 with the Michener family.

Margaret Kelly remembers the excitement the Butlers generated:

> I love to talk about the dudes. It is so nice to be free to say "dudes." Believe me, in those days we didn't dare to call them dudes where they could hear. It was a very exciting thing getting ready for them. Mrs. Butler was of a very nervous temperament. We had to keep the chickens away from her, and we had to guard her in every way. She wanted to ride horseback and wanted my father close at hand at all times to take care of her. But she did make many trips into the mountains.

Julius and Paul Butler with badger they trapped.

> The Butler boys were full of life and deviltry and did many things to torment their mother. One time they told her that a hen had laid an egg in her cabin and they put an egg on her pillow and put a chicken in the cabin, but it was a rooster they had put in the cabin and that caused a lot of disturbance.

The Butler boys at times locked their mother in the outhouse or pelted it with stones while she was in it.

On Saturday night the boys hauled water from the river and heated it in a galvanized tub on the wood-burning stove. They then took their weekly bath in the kitchen of the Michener Camp.

In 1940 Julius Butler told Dr. Caroline McGill about Tom Michener's expertise in throwing a tomahawk. Butler said, "Michener could throw one about fifty feet and hit a circle as big as a fifty cent piece every time. He had one which he had made out of a 7/8-inch rock drill. The blades were about four or five by two inches, just like a double bladed axe." Butler also recounted a story that Michener had once wounded an elk at two hundred feet, and as he approached the elk, it charged him. Michener waited until the elk was close, then threw the tomahawk, splitting the elk's skull wide open and killing it instantly.

Michener made money in his first year as a dude ranch operator, and he planned to go on with his camp. He realized that competition could

Sam Wilson driving a bobsled.

cripple a fledgling venture, so one cold night in December 1906 Michener invited Sam Wilson and Pete Karst to his home. The three men sat round the wood-burning stove and decided to cooperate in developing the dude business. They set a uniform fee for their services: $12 per week for room and board and $6 for a horse and saddle. They called the three dude ranches Michener Camp, Karst Camp, and the Buffalo Horn Resort.

Margaret Kelly remembers that the proposed operation took a large investment that caused hardships for the cash-poor ranchers:

> They used homemade bedsteads, but springs and mattresses had to be bought. Saddles had to be bought. Horses had to be gentled for these unknown eastern dudes. . . .
>
> The service and attention given them by the ranchers and their families as well as by the hired help was unusual and pleasing to these first tourists. In most cases they came with little or no knowledge of the mountains. They came seeking adventure and to see what the Wild West really was like; they came for the solitude and restfulness of the mountains.

Dude ranchers went to great lengths to entertain their guests. The dudes rode, hiked, panned for gold, fished, and collected wildflowers and

Mrs. Jones with trout.

Mesdames Butler and Jones in buggy with Mr. Butler on horseback in the West Fork drainage.

rock specimens. The host families organized fish frys to Porcupine Creek and cookouts at Ousel Falls, complete with chuck wagon. In the evening the guests sat around the campfire and told stories, sang songs, or read the poetry they had written in the mountains. They often toasted marshmallows.

Every dude rancher had to repeatedly answer the same questions of the guests, so the ranchers developed a few tall tales to relieve the tedium of their replies. Pete Karst told his visitors that the Stellar's Jay was a "Karst Blue Robin." Tom Michener warned guests that the only dangerous animal in the canyon was the "side hill gouger," a badgerlike animal with legs on one side shorter than the legs on the other. The animal could only go one way around the hill. If it tried to go around the hill with the short legs on the downhill side, it would tumble off. Michener advised guests to stand their ground and jump over the animal if it came toward them. Cruse Black of the Elkhorn, a dude ranch started in the 1920s, explained to youngsters that the large boulders they saw were brought by the glacier. When asked where the glacier had gone, he replied, "back for more rocks."

One evening Bill Darrett, a taxidermist who preferred to walk rather than ride a horse, came in to Michener Camp. When asked by the young guests where his horse was, he told them it had been stolen by horse thieves. Tom Michener fell in with the joke. He saddled up the horses, armed himself with several guns, and rode up the canyon with all the young people from camp in pursuit of the thieves. With whoops and hollers Michener stopped frequently to have the youngsters hide behind rocks or fallen trees, while he shot at the "bandits." When they finally returned to the ranch the youngsters had seen and counted the horse thieves and knew the number of animals they had stolen.

Mr. and Mrs. Lamme of Bozeman at Michener Camp.

Michener Camp was first to advertise a canyon dude ranch nationally, in a 1907 edition of *Field and Stream*. The ad showed two pictures: one of Lone Mountain and the other of the cabins at the camp. To lure tourists, the ads mentioned the hunting and fishing, wagon and pack trips, the views, and the relaxing atmosphere and promised, "If you wish to hunt, fish or trap, or if you wish simply a quiet place to rest, come to Michener's Camp. We will furnish you with cabins or tents, good beds, good substantial food, gentle horses to ride or drive."

Michener Camp and the Buffalo Horn Ranch took in guests who came to spend six to eight weeks of riding and fishing. Karst Camp, from its beginning, targeted short-term guests, which was understandable because it had long been a stage stop where travelers could get a meal and spend the night. Karst got the bulk of the one-night tourist trade from people traveling the canyon to Yellowstone National Park in the vehicles he provided. Pete Karst, a born entertainer, spun endless tales of adventure and daring in the canyon. Karst Camp dudes had a bar, gas station, museum, and store. In later years the camp boasted a swimming pool, a ski jump, a track where greyhounds raced, and a golf course.

Karst said of his early operation:

> By 1907 I had built several cottages, but there was nobody to occupy them except the forest rangers, and I finally advertised in eastern magazines and got eastern people out here. That was really the start of Karst's Camp—with the Bozeman college professors who would take their vacations up here. Only they didn't call them vacations in those days, they just took some time off.
>
> My earliest guests came from big cities such as Milwaukee. They would come out on vacations and for fishing. But advertising was slow

to take effect—hard to do advertising in those days—hard to tell people what we had. . . . It was very seldom anybody even had a Kodak then, hard to advertise something like this without pictures. My first good customer was C. F. High from Busaris, Ohio. Later he homesteaded here. I advertised in seven different magazines and just those small advertisements cost me about $200 a month.

The Northern Pacific Railroad joined in the advertising campaign and did what it could to lure tourists, who inevitably arrived by train. By 1909 Karst Camp hosted 183 guests, and in 1910 this figure jumped to 600 guests. Pete must have done something right.

Kate Cope, the first school teacher in the basin, lived with the Micheners. While not a dude, she participated in many activities that visitors do during a summer in the mountains. She left an account of some of her experiences:

I arrived there just in the mushroom season, and how Mrs. Michener and I did scour those woods for mushrooms. I was fortunate enough to find a variety new to them, the large black ones, and that made three different species which were edible, and I don't know how many kinds we found of which we knew nothing.

I just had to learn to ride horseback. Why, there were so many places which could not be reached any other way. Yes, it took a good deal of coaxing but after a while I ventured out, and it wasn't but a few days until I could take a ride of fifteen miles.

And some of those rides we did take. They were sure dandy. I must tell you about one of them. You know Mrs. Michener's two sisters spent the summer with her—jolly good girls—and one afternoon when I came home from school, Anna Belle and I decided to go back to the Licks in the hopes of seeing some elk. Tom went with us and you know there isn't a better guide in the West than he. Knows every badger hole in the mountains and can see a grouse a mile away. Why, he took us over places where I had no idea a human being could go, but with him I wasn't a bit afraid.

After we left the trail Tom said we must keep close together and must not talk. Well, you hardly wanted to talk, it was so beautiful. Winding in and out among the pines and across the grassy parks, crossing the patches of snow which still defied the sun, and going on from one slope to another, still higher, until all at once we came out upon a clearing and found ourselves gazing down upon one of the most magnificent grassy slopes imaginable. I saw Cousin Tom drop from the saddle. Anna Belle and I did the same. Creeping up to Tom we looked down the mountain side about one-half mile below us, and there were two large elk, their heads high up in the air and their immense [antlers] glistening in the setting sun. It was surely a sight worth seeing. But the elk had scented us and made a mad dash for the woods on the other side. We tied our horses and started after them, in hopes

> that the timber would stop them and we could get a view at closer range.
>
> How we shot down that mountain. I have never understood how we did it. Fairly leaped along; waded the swamps and jumped over the stones, but we were too late. After tracking the elk about a mile, we gave it up and retraced our steps. The climb back to the horses was hard. It was then dusk and the eight-mile ride home was by moonlight. But the horses were sure-footed and willing, and it seemed no time at all 'til we were at home, half-starved and tired, but happy.

To ensure successful fishing for the dudes, Tom Michener went to the fish hatchery north of Bozeman and got a supply of fingerlings to plant in Porcupine and Beaver creeks. Margaret Michener helped plant the fish. She explains:

> I was about nine years old when I had the opportunity to participate in the first planting of fish in the Gallatin River. My father had worked hard to get fish planted in the river. He finally got the fish hatchery at Bridger [Canyon] to provide the fish and he sent his hired man, Clyde Tedrick, to bring the first fish up. There were about six big milk cans full of little trout, and we understood that we had to keep them aerated—we had to keep them jiggling. I was delighted to be the one to ride along with those fish. He [Clyde] would drive at a trot, and the rough roads kept those fish well stirred up. We thought we had to add fresh water every few miles. It was a lot of fun to grab the pail and run for water, but before half the day was gone it was a mighty hard thing to think of carrying two pails to each can while Clyde would stir around among them and throw out any fish that had died.

Mrs. Johnson and Maggie Michener in fancy western dress.

The 320 Ranch in early 1930's. —U.S.D.A. Forest Service

I was a weary child when finally my father met us. Then my father put in a faster team and didn't think it was necessary to make so many stops. I'll never forget how sad I was when my brother and my father took one-half of the fish to Beaver Creek and Clyde took the other half to Porcupine, and I was left behind. The little tiny fish were dumped into the creeks, and many people in the canyon said they would never grow, but it wasn't long before the brook trout were being caught in both creeks.

In 1919 the Orlando Russell Family of Grand Rapids, Michigan, made a trip through Yellowstone National Park. The rough roads made travel difficult, and they covered less distance than they had planned. They found themselves exiting the northwest corner of the park in the early evening:

The park ranger told us Wilson's [Buffalo Horn] Ranch [now the 320] eight miles further could accommodate us for the night. (It was 6 o'clock and we had not made as far as expected owing to the rough roads.) Wilsons proved to be a fascinating looking place to a tenderfoot, large log house, with half a dozen small cabins around. We secured a cabin for the night. We are on the bank of Gallatin Creek, (or river). Boys fishing—The Rockies all around us—[canyon walls] rising from the rear—sunken forest nearby. Most delicious supper—fried potatoes, cold roast beef, jelly, peas, fine milk, hot biscuits, cake, rice pudding, coffee.

Mr. and Mrs. DeWolf cabin next to ours—wheat man from Minneapolis, twenty-acre home on lake Minnetonka . . .

After supper a prowling squirrel commenced to chatter in a nearby tree. Mrs. Wilson stepped into the house, brought out a gun, took quick aim and the little squirrel "ceased to chatter." Another squirrel was dispatched the same way, just casually. Some shot! Boys fishing. Sleep? Well, yes.

July 21st. Breakfast 8 A.M., oatmeal and thick cream, fried potatoes, ham, hot biscuits, marmalade, syrup, coffee, milk. Fine. After, boys went fishing. D. M. R. and self went for mountain climb. It didn't look terrific but we climbed over one height to a valley—one after another. When we were ready to take the last hop I told D. not to wait for me. Soon he was out of sight behind an aspen grove. I stretched out in a spot of shade and looked away and away—straight down into innumerable defiles winding through mt. after mt. and up into a blue sky dotted with billowy clouds. After a bit, shrill screams and I saw a great bird with wide spread wings circling away off over a peak. Nearer and nearer and circling down quite close came a white-headed eagle screaming as he came, so close, yet he did not light.

Back to the foot of the mts. 12:50. Stopped on Gallatin River, took off our shoes and stockings and both sat on a log out in the water and paddled in the springfed coolness of the stream. . . . After a bit our boys were helped into chaps, all with guns. . . . Excitement! They drove cows, went up the valley. . . . Had horses nearly all day. Mrs. Wilson had waited dinner until we came down from our climb. Fried chicken, rice, mashed potatoes, gravy, corn, fresh bread, homemade butter, coffee, milk (and such milk), pudding. Slept a bit in P.M. Very hot here from 3-6 P.M. until dark—fire in the morning, cold until near noon. Supper at Wilson Ranch—mountain trout, cream cheese (great), fried potatoes, strawberry shortcake and whipped cream, buttermilk such as we never drank before. . . .

Mrs. Wilson remarkable woman. Once a teacher, now member of Board of Education, postmistress, remarkable cook and housekeeper, crackshot. (Many deer and elk to her credit) No school in twelve miles. Last winter she felt sorry for the children around and taught.

In 1919 the Micheners moved from the canyon, ending their dude operation. In the 1920s the Lemons at Half Way Inn on Twin Cabins Creek entered the dude ranching business to take advantage of the traffic to Yellowstone. Besides keeping a few guests the Lemons ran a store and gas station.

In 1922 Grace and Ernest Miller started the Elkhorn Ranch almost by accident. Pete Karst had promised the newly married, young couple jobs as wrangler and waitress. Grace Miller had gone east and lined up four dudes to stay at Karst Camp. Then Pete Karst broke the news to Ernest Miller that business was down. Pete offered Ernest a job but could not give Grace employment. Further, he could not give the Millers a 10 percent commission on the four dudes that Grace was bringing west with her. As Grace tells the story:

Early Elkhorn. —U.S.D.A. Forest Service

Ernest . . . went over to the Elks club and was having a quiet beer at the bar feeling pretty forlorn. Suddenly the . . . door opened and in stormed a man who was perfectly furious.

It seems . . . he and his wife the summer before had had a chance to buy for tax title a delightful five-acre location way up the Gallatin right next to the park with a lovely little cabin and barn. The cabin had two bedrooms, a sitting room, and a kitchen with a couple of porches. . . . Incidentally, it had two big piles of beautiful elkhorns piled up each side of the front door. . . .

Alas, they'd had no experience with cabins in the mountains and had only locked the door when they'd left, leaving everything ready for their return the next summer including food stuffs neatly in containers on the pantry shelf. He had found the cabin in a horrible mess. A grizzly bear had broken the window and gotten in, knocked all the food off the shelves, had slobbered some chocolate and honey all over the beautiful bedding, and the rats and other animals had come in after that, and what the bear hadn't ruined, they ruined the rest.

He spent a long time cleaning up the mess as best he could and boarding up the cabin windows and started back to town. It had rained, and it was slippery, and his car slid over against a nasty rock and broke its rear axle. Ernest knew that whole country and knew this particular homestead very well. And it was just exactly down our alley.

He [Ernest] quietly worked up the bar to where the man was sitting and quietly questioned, "You wouldn't be interested in selling that place, would you?" The man turned and said, "You're damn right."

Ernest said "How much?" The man replied, "Five hundred dollars." Ernest took two fifties that he had in his pocket and handed them over to the man and said,"I'll bring the rest tomorrow."

The dudes spent their summer in the quaint log cabin, and the Millers slept under a tree.

Dude ranches belonged to a state association that received many job applications from eastern college students anxious to experience the Rocky Mountains. Ranchers got many waitresses and housekeepers this way, but eastern students did not fill the bill for wranglers and choremen. Grace Miller went to Ennis one day to see if she could find a good choreman.

In those days, there was one bar in particular in Ennis run by one of the Clark boys that was sort of headquarters for choremen and cowboys. When they [felt like] a semi-annual spree they would go down to Oscar's Silver Dollar saloon and Oscar would look after them and was pretty good at sobering them up when the time came and was good to them in his way. Of course, by then they had used up most of their summer wages and were looking for a job. We always went there when we needed extra help.

Grace Miller, with her friendly personality and love of people, perfected the art of going east and lining up dudes for the summer. She departed from the method used by the early dude ranch operators who relied on magazine advertisements, help from the Northern Pacific Railroad, and word of mouth.

Julia Bennett, who worked at the Rising Sun, later named the Nine Quarter Circle, left an account of how she went to New York City in the late 1920s to book guests for her ranch. By then the cost of a summer vacation in the mountains had increased considerably. In spite of high fees, the dude ranchers still operated on a poor cash flow. Julia Bennett visited a woman in New Jersey who asked her if she required a deposit.

But I said, "Oh, no." I went back to my little apartment hoping to hear from her. At the end of three days she called and said they had decided to come. There would be two adults and one child for the summer. The next day I received a check for $1,000. It was a $5,000 booking. . . .

Conditions were just as bad in 1932, but ranchers were going to New York to book . . . they took window exhibits. . . . I got an exhibit together and . . . it consisted of a buffalo skin, sage brush, saddles, chaps, silver mounted spurs and bridles, a six shooter, and an old buffalo skull. I . . . took my exhibit to the Northern Pacific Office. They placed it in the front window on Fifth Avenue and it surely attracted much attention. At night a large crowd would gather around and talk about it. I used to mingle with the crowd and listen.

In 1925 Vic and Eda Benson started the Covered Wagon Ranch just below Taylor Fork. Vic Benson came to Montana from Minneapolis.

After spending time in Bozeman they decided to open a guest ranch in the canyon. They chose their site for the view and named it the Gallatin Way Ranch Company. Eventually they called their camp The Covered Wagon, because they thought the name more appropriate and because Vic Benson's brother had a restaurant in Minneapolis named The Covered Wagon. Hunters, who paid $50 a week to stay with them, were their first patrons, and their summer business developed by word of mouth from the hunters.

In the 1920s the Butler family, which had spent many summers in the Canyon, stayed at the Rising Sun Ranch on Taylor Fork. Julia Bennett, who worked for ranch owner Seymour Cunningham tells of the preparations she made for the family: "In the spring of 1920 Mr. and Mrs. Fred Butler and Mr. and Mrs. Don Kilbourne [their daughter and son-in-law] rented the Nine Quarter Circle Ranch on Taylor Fork to spend their summer vacation. They hired a crew to repair cabins and put in water, toilets and showers. They hired me to cook for the crew and fix up cabins. I bought new furniture for the cabins, put water in the kitchen sink, new range, painted floors, new rugs and new dishes." The following year the Butlers returned. Bennett continues, "The next summer they came to spend their vacation. They wrote me to hire them a cook, so I hired myself."

In 1927 Paul Butler bought the Adams homestead behind the Elkhorn Ranch on Sage Creek and built a compound, which he called the Seven-Eleven Ranch. Butler entertained his own guests at the ranch instead of using it for a commercial operation. He raised polo ponies at the Seven-Eleven and shipped the horses to Chicago for training.

In the same year, his uncle J. Fred Butler tried to get The Rising Sun Ranch from Seymour Cunningham. The cattle partnership of Cunningham and Behring had gone bankrupt in 1920, and after restructuring the partnership dissolved in 1923. By 1927 Seymour Cunningham needed to clear title to the ranch by paying the back taxes. J. Fred Butler paid the back taxes and prepared to take possession of the ranch. But before J. Fred could get title, his nephew Julius Butler stepped in and loaned Cunningham the money to clear other debt on the property. J. Fred lost the tax money he had put up. He then went to the West Fork where he bought the Clarence Lytle homestead for the unheard of price of $50 an acre and began the B Bar K Ranch, which today is the Lone Mountain Ranch. The Butler and Kilbourne families used the B Bar K Ranch for their own recreation, rather than as a paying dude ranch.

Meanwhile, back on Taylor Fork, the Cunningham family, first Seymour and then about 1940 his son Bob, ran the Rising Sun Ranch as a dude operation. The ranch enticed film star Gary Cooper, a native

Montanan, to spend time on Taylor Fork during the 1930s. Clara Bow also stayed at the ranch during the depression, which provided good publicity. Besides bringing in movie stars, the Cunninghams brought in top rodeo riders to provide entertainment for their guests.

Dude ranching reached new heights at the Rising Sun when the Cunninghams brought in a French decorator to redo the ranch. The ranch's main lodge had a large fireplace with seats built on either side of the opening so the guests could warm themselves by the fire. The decorator used clever western motifs, including deer feet to hold up curtain rods and elk antlers as door handles.

During World War II Bob Cunningham stopped operating the ranch, whose name he changed from the Rising Sun to the Nine Quarter Circle. The buildings fell into disrepair, and at war's end Howard Kelsey bought the ranch, which his family still operates today.

Pete Karst sold out to Harold and Esther Carlson in 1951. The dining room at Karst Camp burned to the ground in 1957. The Carlsons rebuilt and eventually sold the ranch. A second fire leveled the log restaurant in the late 1970s. The new owners rebuilt on the east side of the highway, away from the river. This building, too, burned in the 1980s. It has never been rebuilt.

Until the 1980s the Buffalo Horn Ranch continued to operate as a dude ranch.

The Wilsons sold the Buffalo Horn Ranch in 1926 to Luke Brown, Pete Kellorn and several other investors for $32,000. Brown paid the Wilsons $11,000, but the rest of the group failed to come up with the money to complete the sale. Sam died in 1929, and Pete Kellorn managed the ranch for Josie Wilson until she sold the property in 1936 to Dr. Caroline McGill, after which Park and Susie Taylor managed the ranch for Dr. McGill for many years. After the Second World War, Patty and Jimmy Goodrich managed the ranch. When Dr. McGill died, they bought the operation. Now called the 320, the ranch operated continuously from 1907 into the 1980s, longer than any other ranch in the canyon.

After the war people became more mobile, spending less time at dude ranches and more time touring the West. Others learned of the adventure in pack trips, back-country tours with guides who put up tents, cooked meals, and wrangled the horses. Kay and Tommy Fisher led pack trips for the 320. Luella and Walter Latta organized the pack trips for Karst Camp. For the first two nights the campers ate fresh steak and salad. After that the cook served frozen meat, and by the end of the week the campers ate canned ham and chicken.

Spanish Creek

Soon after the founding of Bozeman, Zachary Sales brought his family from Ontario, Canada, to the Gallatin Valley. Sales located south of Bozeman and built a sawmill, digging a holding pond and mill race from the Gallatin River to the mill. The town of Salesville grew up around the sawmill, and residents of Salesville soon ventured south into the canyon where they cut timber and grazed cattle.

The abundant fields of wild hay in the area, good for grazing and cutting, impressed the farmers who ventured south to Spanish Creek. Homesteaders slowly filtered beyond the mouth of the canyon into the area. The Thomas Lemon family, from Missouri, was one of the first to settle at Spanish Creek. In 1882 Lemon moved his family from Salesville to a one-room cabin at the mouth of Spanish Creek. The family moved into the cabin, with dirt floor and dirt roof, prior to installing the windows. Before cold weather arrived they put in the glass, making the house snug for the winter.

Lemon grazed cattle and horses, raised small grain, and cut hay as he expanded his ranch over the years. When cattlemen began running their herds into the canyon, the Lemons provided meals for the herders as a way to bring in cash.

In the 1880s valley residents considered the Spanish Creek area remote. Lemon explained his reason for settling in the area: "There was plenty of water and grass there and it would be pretty hard to starve him (sic) out there as these were the things that everyone had to have." The remoteness still caused some problems—he had to wait until the harvesting crews finished threshing at Salesville before he could get Sam Krattcer to transport his horse-drawn equipment up to Spanish Creek to thresh the grain that grew there.

Through the end of the nineteenth century homesteaders continued to move into the Spanish Creek area. W. J. Bradfield established a sawmill in the Spanish Creek drainage to fill his and the other settlers' need for timber. In 1901 Joachim Kundert received permission to operate the Spanish Creek post office from his home near the divide between Spanish Creek and the Cherry Creek Basin on the Madison side. Kundert brought the mail from Salesville three times a week and saved the settlers a trip to town.

As the number of homesteaders increased, the county built several schools. Like the schools in other sparsely populated areas, the Spanish Creek schools operated when they had enough students. School often closed down in the winter when the children could not get to the school house.

Men far outnumbered women in the West. While some male homesteaders married and raised families, many others remained bachelors. A few men got brides from catalog lists of women who wanted to come out west and marry a rancher. Some of these "mail order" marriages turned out happily, but at least two at Spanish Creek ended tragically. Both women arrived by train, immediately married in Bozeman, and rode to their new homes in lumber wagons. They lived in crude cabins far from civilization. The next year both brides died in childbirth, without benefit of a doctor and without ever having been off the homestead.

By 1900 homesteaders had claimed most of the public lands in the Spanish Creek area. As the years went along, many settlers found that 160 acres was too little land to provide a living and so they opted to sell their homesteads and move on. W. W. Wylie, who owned the Wylie Way Camps in Yellowstone National Park, began to buy up Spanish Creek homesteads. Over the course of several years Wylie consolidated his holdings by buying eleven sections of land on Cherry and Elk creeks on the Madison side of the divide between Spanish Creek and Cherry Creek, and eight sections on Spanish Creek in Gallatin County. The Thomas Lemon Ranch at the mouth of Spanish Creek went to Wylie.

Wylie used the land to winter the three thousand horses that pulled stagecoaches filled with tourists and hauled supplies to his camps in Yellowstone. Wylie and his family had come from Kansas to Bozeman, where Wylie served as superintendent of schools. He quickly realized the tourist potential of Yellowstone and set up a string of tent camps to provide a complete vacation experience to the tourists flocking to the Rocky Mountains. In addition to wintering horses, the cowboys at the Spanish Creek Ranch broke the young stock to work in harness and for trail riding. Wylie also raised sheep on Spanish Creek. Wylie's sister, Belle Lockhart and her two sons, Wylie and Eaton Lockhart, ran the ranch.

In the spring, wranglers and cowboys drove three thousand head of horses south from the Spanish Creek Ranch, through the Gallatin Canyon, and over the Big Horn Pass to Mamouth Hot Springs in Yellowstone Park. Margaret Michener, who lived at West Fork, loved to see the sleek, fat, perky horses going to the park but disliked watching them pass on their way down the canyon. The horses returned in the fall, thin, lame, and covered with saddle sores.

In 1906 Wylie and Belle Lockhart opened a summer camp at Spanish Creek similar to their operation in the park. A four-horse Concord wagon picked up the campers at Bozeman and transported them to the ranch at Spanish Creek. Wylie and Lockhart charged $3 each way, and $12 a week for room, board, and horse, or $2 a day. This dude operation only lasted one year, but the Spanish Creek Ranch continued to provide a natural stopping place for travelers going into the canyon. Many local residents spent the night or had a meal at the ranch before resuming their trips.

The first large social event in Spanish Creek, the marriage of the Lockhart's youngest daughter, took place in September 1906, the first of many glittering events to take place, over the years, on the rich and profitable ranch. The Lockharts erected a large tent, which they decorated with green boughs and red berries and a wagon load of sweet peas sent especially from Bozeman. The bride and bridesmaids wore white dresses with lace insets. The groom wore a black broadcloth vest with English gray striped pants.

While Wylie ran his ranch on Spanish Creek, the Anceney family moved into the area. Charles Anceney, Sr., and his wife Marie Angelique, immigrants from France, arrived in Montana from Missouri via the goldfields of Colorado. Like many early settlers they came because of the mineral wealth found in the West; unlike many others the Anceneys brought cattle with them. After trying their luck in the goldfields of Virginia City and Nevada City, the family settled in 1865 near Manhattan in the Gallatin Valley where they established a cattle ranch.

In the 1880s, in search of pasture, they began to summer their herds in the Gallatin Canyon as far south as the upper basin. A bad winter and some worse luck with a debt left the Anceneys broke. They borrowed $38,000 from a bank in Bozeman, and with this money they bought cattle in Nevada and moved them onto Northern Pacific lands on the Madison side of the divide from Spanish Creek. By 1894 the Anceneys owned or leased 16,000 acres on both the Madison side of the mountains and on the Gallatin side in the Spanish Creek area. Father and son operated a cow-calf ranch and also raised purebred Shorthorn cattle. Disaster struck in 1895 when a saddle horse fell on Charles Anceney, Sr., pushing the saddle horn into his chest, leaving Charles Anceney, Jr., to carry on the cattle business alone.

Charles Leon Anceney, Sr. and Jr. —C.L. Anceney III

In 1907 W. W. Wylie sold his ranch on Cherry Creek to Harry Child, a Helena banker who owned hotels and the transportation system in Yellowstone Park. Around 1907 Anceney, Jr., looking for a source of cash, began working with Child. Anceney, Jr., proposed that he winter Child's one thousand park horses on his lands at Spanish Creek. By returning healthy horses to Child in the spring, Anceney proved his worth. Anceney continued to run his own herds of cattle and winter Child's horses for several years.

By 1911 the Anceney cattle operation at Spanish Creek became so lucrative that the Northern Pacific Railroad built a line from Manhattan to Amsterdam and then south to the boundary of the Flying D Ranch, giving the cattle operation ideal transport. The railroad named the town at the railhead Anceney.

In 1912 Child and Anceney realized that automobiles and buses would soon replace all the horses used to carry tourists through Yellowstone Park. The two men decided to stop pasturing horses and switch exclusively to raising cattle. During 1913-14 Anceney and Child entered into a partnership and bought up additional acreage in the Spanish Creek area. Anceney, Jr., put up the lands that he owned or leased, and Child provided the cash to increase their acreage. They called their ranch the Flying D.

Charles (Chuck) Anceney III explains how the ranch came to use the "Flying D" brand:

> While working in the blacksmith shop at the Home Ranch, he [Anceney, Jr.] fashioned a brand and holding it up asked his foreman, Ernest Vaughn, "What is this brand?"
>
> Flying U
>
> Ernest answered, "The Flying U." Father said, "There is already a Flying U" and turning it over he asked, "What is this?" Ernest replied, "That is the Flying D."
>
> Flying D
>
> Father then said, "That is the new brand of the outfit you work for." To adopt the brand Father had to convince Mr. Child that it was a natural. So he explained it to him as a C for Child and a D for Dean, Mrs. Child's maiden name. Mr. Child thought it was a great idea.

To secure the cooperation of the railroads that served the area, Harry Child induced several officers of the railroad to enter the partnership. Included in the group, with a 23 percent interest, were Charles Perkins, a director of the Northern Pacific Railroad, Howard Elliott, president of the Northern Pacific, and a Mr. Scott of the railroad.

By 1914 the land controlled by the partnership encompassed ten townships, five school districts, and two churches on both the Madison and Gallatin sides of the divide. The ranch bought up approximately 110 homesteads and combined them into rental ranches operated by the tenants. Eventually the ranch encompassed 100,000 acres of deeded land and 400,000 acres of leased land. It measured 24 miles wide by 26 miles long, and over 550 miles of fence enclosed the property. The range supported eight thousand to ten thousand head of cattle bought at about $30 a head. The partnership had purchased the land for about $10 an acre.

During the First World War, Child plowed up thousands of acres of hay fields to plant wheat. Anceney, Jr., objected strenuously to plowing up good range land. The ranch used an enormous steam engine with twelve-foot wheels for plowing. The operator needed an engineer's license to run the huge machine, which used gang plows, ten across and several deep. The crop came in as the war ended, removing the market for the ranch wheat. This grain-growing attempt created a large debt for the ranch and ruined thousands of acres of grassland.

About thirty-six of the original homesteaders stayed on as tenant farmers and raised hay for the Flying D. These tenants included Bill McCloud, Berg Clark, Merle Alexander, and Rutledge M. Hargroves. Chuck Anceney recounts how the system worked:

> There were also some cereal crops grown, and lots of hay. Although the numbers of stock were greatly reduced during the winter, most of the

hay was used on the ranch. The producers were paid a straight $6 per ton fed out. If the market price was higher than $6 per ton, the renter still got $6, but if the price was lower, they were always paid $6. Usually the price was below $6 and everyone was satisfied.

The ranch supplied the tenant farmers with mowers and rakes.

Anceney, Jr., had always maintained an interest he inherited from his father in purebred cattle. While Anceney, Jr., operated the Flying D, he continued the family interest in blooded stock. The Flying D kept a herd of purebred Shorthorns. These cattle from England gave more milk and had a wider loin, hence more preferred cuts of meat than typical Montana range cattle. In addition the cattle were good rustlers and survived on the rugged range in Montana. The ranch also used crosses for hybrid vigor. The calves of a Shorthorn bull and Hereford cow made for a particularly sturdy strain. Anceney, Jr., also kept a show bull named "Ringmaster," who won the title of grand champion three times at the Chicago International Fair. Archie Kane managed the show cattle for Anceney.

Charles Anceney, Jr., Harry Child and Cumberland Sire, a prize winning bull of the Flying D Ranch. —C.L. Anceney, III

Usually the ranch operated a fat steer operation. Anceney had originally started in the cow-calf business, called a wet ranch. The economics of ranching in Montana dictated that the ranch fatten up steers for sale at the Chicago market, so the Flying D bought cattle in Texas and Mexico, put them up for three years, and then shipped them off to Chicago.

During the early summer, ranch work required cowboys to be up at 2 A.M. They would herd cattle until 10 A.M. and then retire to the home ranch where they would sleep and get two good meals. At 5 P.M. the cowboys would go out and herd again until about 10 P.M. This routine lasted six weeks.

Chuck Anceney, born in 1914, was a cowboy from age five when his parents gave him a horse, which he rode around the corral all day. Chuck recounts that, "One day I opened the gate and rode up to Dad and asked, 'Do you need a good hand here?' Dad said, 'Yup, you go over there and hold the cut.'" From then on Chuck rode every day in the summer.

During the 1920s and 1930s the Flying D shipped cattle to Chicago four times a year. Each shipment consisted of thirty-six to forty railroad cars with each car carrying thirty-two head of cattle. The Anceneys, father and son, made the last trip to Chicago in 1936. Chuck tells about shipping the cattle to market:

> One of the most skillful and crucial phases of the whole business was marketing the cattle. First was the choice of time of arrival in Chicago, hoping to catch the market up one cent rather than down two. The cattle were driven in a leisurely manner to the Anceney corrals, with the hope of picking up a last several pounds rather than losing a good many in the inevitable confusion which came with the drive.
>
> The cattle were then loaded, and after endless and persistent negotiation, the Northern Pacific Railroad was persuaded to provide almost non-stop transportation. While in motion the cattle remained quiet, engaged in the task of keeping their balance. On even a short stop they would try to mill around, becoming disturbed and hot, resulting in loss of weight. The railroad finally worked out a fast freight schedule of about sixty-one hours from the ranch to the Chicago yards. Humane and legal considerations required an overnight lay-over at Aberdeen, South Dakota, where the cattle were unloaded for rest, feed, and water.

Ray Michener started working on haying crews in 1923 at the age of eleven. He continued to work in the Spanish Creek area until he graduated from Montana State College. He gives a detailed account of hay cutting and the grain harvest at the farms rented by the tenants of the Flying D. Tenant farmers, itinerant workers, and college students with experience worked as a crew going from farm to farm during the season. The first crop of alfalfa hay was followed by grass hay and then the small grain grown in the Spanish Creek area. When the crew finished harvesting these three crops, it would bring in the second cutting of alfalfa hay.

Michener writes:

> The first job was to get the horses in and reintroduce them to harness. It was like breaking them all over again. A wild horse would be hooked up with a gentle one and put to hauling a stone boat, a sort of sled, flat on the ground, loaded with rocks. It was just a drag that the horses couldn't run with. After the wild one quieted down it was turned over to the mower crew.
>
> The first haying operation was to cut the hay. It was a matter of pride for the mower to cut a field cleanly. Care was taken at turns to leave no grass standing. It was grass while standing, hay when cut. . . . A good mower team could cut from seven to ten acres per day. Hargroves sometimes operated as many as fifteen mowers at one time. McCloud operated four to five.
>
> The maintenance of the mower equipment was an important task. Cycle blades had to be sharpened, worn out cycles replaced. Hargroves had to change cycles twice a day on account of rodents building piles of dirt in the grass. . . . McCloud had his mowers sharpen their cycles at night.
>
> After the hay had dried the correct amount it was raked into windrows. The correct amount of moisture was determined by twisting a handful until it broke: two turns, pretty dry - three turns, about right - four turns, too wet. You couldn't do anything about too dry hay. With the too wet you waited until it dried. . . .
>
> [The hay] was raked with a sulky rake - a two-wheeled rig ten feet wide. The wheels . . . carried an axle between them on which some thirty curved steel rake teeth were mounted. When the operator stepped on a pedal a nub engaged the gears, dumping the hay from the teeth into windrows. Mounted on the frame was the operators seat, directly above the rake. . . . I would figure on raking thirty acres per day. McCloud would have two to three rakes on the job. Hargroves kept three to five rakes going.
>
> The rakes were light and, if a team decided to run, it was impossible to hold them. . . . When they ran you were in imminent danger of being thrown off your perch, down back of the horses and in the rake. You would then get rolled around and around until the rake hit a bump or a ditch that caused it to dump. . . . I developed a method . . . for when my team ran. I would just turn over backwards and fall off. I would keep hold of both lines if I could but one for sure. Then I would circle the team, out at the end of the lines until it wore out. It worked real well. You were glad to take that backward tumble rather than be rolled.
>
> Next the hay was stacked. It was brought into the stack with bullrakes . . . sort of a front end loader with long wooden teeth. These

teeth ran along the ground, scooping up the windrowed hay. It was propelled by two horses working about eight feet apart.

Stacking was done differently on the individual tenant farms:

McCloud usually built small (twelve to eighteen ton) stacks. . . . Hargroves used four bull rakes, and make larger (twenty to thirty ton) stacks. . . . The main idea was to get it into stacks that wouldn't fall down. Hay put up as fast as this settled awfully fast and sometimes unevenly. Many the time we had to prop up, with poles, a stack that had been straight when completed. But not to worry. If the stack would hold up a tape for measuring it was okay . . .

The haying crews, numbering up to thirty-five men, worked ten hours a day. They stayed on the farm. They lived in the bunkhouses provided by the farmer. . . . During my high school and college days I worked most of my summers on one Flying DRanch or the other. . . . I went out to work for Bill McCloud when I was about eleven years old. He put me to driving a hay rake. I did well at that job and my second year Bill paid me man's wages— $3 per day vs. the kids $1.50 per day.

I graduated from the rake almost directly to the stack. Stacking was terribly hard work but I wanted the money for college. They paid $5 per day for stacking vs. $3 per day for other work. Those company tenants were putting up hay on a contract and they put it up fast . . . We generally put up four or five stacks a day. You got so you didn't even look back at a finished stack.

The haying season faded into the grain harvest. Here the grain was cut with a binder. . . . The bundles were gathered up by hand and placed in shocks with the heads up. The butt ends of the shocks were socked into the stubble. [This protected the grain from rain or snow damage.] Then the bundles were pitched on to a wagon by the bundle team and hauled to the threshing machine. . . . Bundles were then fed (grain end first) into a conveyor leading into the thresher. The grain was extracted and fed to wagons that hauled it to the granary. The straw was blown into the stack for other uses.

Few of the farmers owned their own threshing machines or had bundle teams for their own crops. All the big outfits were custom jobbers. A farmer might own a thresher but he operated it on a custom basis. I had one of Bill McCloud's bundle teams and wagons. You just took your bedroll with you and traveled with the machine. You slept in the barn wherever you stopped.

The last operation was to feed out the hay in the winter. Here you would load the hay out of the stack onto a rack mounted on a sled and then distribute it in small piles for the cattle. I fed two winters for McCloud. He would give you about six hundred head to feed. It would keep you real busy. . . . A good deal of hay was wasted in trampling.

That wintertime work had its problems. The hole in the river where you watered your team would freeze over and you would have to chop it out. You had to bring the bridles into the bunkhouse at night so that

> the bits were warm. . . . You had to bundle yourself up real good—especially your hands and feet. We generally wore two pairs of cotton gloves with the outer pair turned inside out. . . . For your feet we found that to set your bare foot down on two or three pages of newspaper, wrapping it around your foot and inserting the whole thing into a cloth overshoe worked very well. These precautions were prompted by weather that commonly dropped to 40 degrees below and colder.

Another ranch operation involved mucking out manure and transporting it to build dams in irrigation ditches. Michener considered picking rocks the worst job on the farm. The picker put the rocks on a wooden sled and then delivered them to a rock pile, where he dumped them off. Michener describes rock picking as "lonesome, tiring, boring work."

In 1929 the Flying D partnership became a corporation. In 1931 Harry Child died and his son-in-law, William Nichols, took over his interests. In 1936 Charles Anceney, Jr., died in an automobile accident. The finances of the Flying D had gone through many hard times, with Anceney's share consumed during the depression. With his father's death Chuck Anceney's involvement with the ranch ceased. Nichols continued to operate the ranch until 1944. The Nichols family retained 10,000 acres as the Spanish Creek Ranch and sold the rest of the Flying D land to the Irvine Corporation of California. In the 1950s James Ray bought the Spanish Creek Ranch and 14,000 acres on the Madison. Ray sold all the Spanish Creek Ranch land, including the additional 14,000 acres, to R. Shelton in 1971. In 1978 Shelton consolidated his holdings by purchasing the Flying D Ranch, for a total holding of 130,000 acres of deeded land, about the size of the original ranch. The ranch last changed hands in 1989 when Ted Turner bought the entire acreage from Shelton.

The Legendary Pete Karst

Pete Karst came to the canyon early in 1902 and participated in its many changes until his death in 1966 at age 90. For many tourists and Bozeman residents, Pete *was* the canyon. He loved to sit in his bar and regale friends and strangers with stories of the area, the more colorful and exaggerated the better. Pete tirelessly promoted the canyon to dudes, investors, and anyone who might use his facilities or work his mining claims. Many people heard him exclaim, "The canyon is a beautiful spot. Nothing like it, winter or summer. Always something interesting, and a man lives free."

Karst passed through the Gallatin Valley in 1898 at the age of eighteen, on his way from Wisconsin to the West Coast. He got as far as Lewiston, Idaho, then turned around and headed back to Bozeman. In 1902 Karst landed a job driving freight and loggers from Bozeman to the Cooper Tie Camp, which Walter Cooper had just opened on Taylor Fork. This job required many horses, wagons, and drivers, and with the encouragement of Walter Cooper, Karst got a loan from the bank and bought the necessary equipment to conduct a freight operation.

Pete explained to Dr. Caroline McGill how he chose the place for his overnight stop in the canyon, thirty-five miles from both Bozeman and Taylor Fork:

> This was the spot I picked for the overnight stop—right distance between Bozeman and the tie camp. This was unsurveyed land. I had a partner then and so we operated together for about a year and then I bought him out. His name was Oliver. Bought twenty head of horses from him. He was the squatter on this place and I bought the squatter's rights from him in order to get possession here. Before that, the fellow

Wedding photo of Pete and Jennie Karst. —Sara Anderson

that was the real squatter here had a sawmill dated back to 1889 when he took out a small ditch to irrigate the land over across the river. Those improvements are what I paid $1,000 for at the time: horses, homestead rights, unsurveyed land. Oh, I don't know that it was so good—meant a lot more work—those 160 acres I bought in 1903.

> Big Lew was the name of the fellow who was the real squatter here—the one that had the saw mill. He got the timber out to build a halfway house here. Then the national forest was created and the forestry was undecided whether they would let anybody homestead this land. So that wasn't settled until the survey was made by Bushnell in 1906-1907 from West Fork on down to here. In about 1905 I started building on my own.

Pete ferried workers up and down the canyon as well as delivering supplies and mail to the tie camps. And, for good measure, he provided the tie hackers with liquor, a commodity that Walter Cooper worked hard to exclude from his camps. Pete explains:

> Just to let you know how things would go then—we had no bar here and no way of dispensing liquor, and we had lots of tie hacks and lumber jacks who liked liquor. So I had a little pigeon hole I served the drinks through—think we charged them 50 cents a drink—liquor cost about $2 a gallon. But the tie hacks could only handle a few drinks—there's a difference between a little whoopee and too much. Got to be kind of hard to handle. Finally I got tired and sent a man up to tell these boys that the county was complaining and the county attorney was just a little way down the road. By God, those tie hacks took their bundles and lit out across the country and I never did see them again.
>
> Those tie hacks didn't do so bad. They would make up to $10 or $15 a day. Their checks would run up to several hundred dollars. If they worked for the winter, six months or four months, they'd have a nice stake of $1,500 or $1,800. But except on pay day they never had much money at a time.
>
> The passenger fare, back when I was freighting, was $10 and 5 cents a pound for baggage in excess of twenty-five pounds. The baggage all my passengers had was heavy—broad axes and hooks—every one would have one hundred pounds of excess baggage. Usually had eighteen to twenty people on these rigs besides the baggage—mostly tie hacks, lumberjacks, river drivers.
>
> Then Cooper and Company went broke, owing me quite a lot of money. So there I was—had lots of horses, wagons, sleds—but no money. The Company promised that as soon as they started up again they would take care of the money they owed me, but they never started up again, so I was a sucker that time. . . . I would venture to say that about the time the Cooper Company went broke I was about $30,000 in debt.

The same year that the Cooper tie operation went broke Pete had a bad accident that left him "listing to port" for the rest of his life. The *Republican Courier* of 29 November 1907 gives an account of Pete's mishap:

> Pete Karst, who owns the summer resort in the West Gallatin Canyon and who carries the mail from Bozeman to Eldredge, was mixed up in

a runaway last Friday and sustained injuries that will keep him confined to his home for some time. He hitched up a bronco with another horse last Friday morning, to a wagon which had a very high seat. The team ran away right in front of his home . . . and when running at top speed struck a large boulder, throwing Mr. Karst from his high seat into the rocks and boulders by the side of the road. When picked up he was unconscious and an examination showed that both his upper and lower jaws were badly fractured, his right arm broken and his left wrist sprained and fractured.

Dr. Patterson was sent for and upon his arrival . . . did everything possible to make his patient easy. . . . While the shock to his system is severe, it is believed he will pull through as he has a good constitution.

Pete continued to haul freight and mail into the canyon while he looked around for another way to make money. In 1906 Karst, Sam Wilson, and Tom Michener entered into an agreement to start dude ranching in the canyon. This allowed Pete to use his freight wagons to transport tourists up and down the canyon. Pete kept many of these visitors in the cabins he had built at his place. Kate Cope, the first school teacher in the basin, has left an account of her trip in 1908, via Karst Stage, to the Michener Camp:

How I wish you could have been on that first stage ride. Now don't get it into your head that it was one of those handsome yellow Concord stages with red seats, drawn by four prancing grays, the driver one of those old-fashioned stage drivers who rules his life just so, and cracks his whip just so, and tells you all kinds of experiences as you sit up beside him—the seat of honor, you know. Oh no, my dear. This stage may one day have boasted yellow paint and red seats but now was a very humble old road wagon with a white canvas top and what few bolts were left were tied up with string and rope. Instead of prancing grays the wagon was pulled by one poor little bay and its mate a sorrel, unshod and poorly fed, as willing as could be.

I shall never forget how that little team pulled, how bravely they worked until they could work no more. And there was no need for the driver to hold the reins or crack the whip through those muddy roads. You know that the basin is sixty miles from Bozeman and it rained every step of the way, except of course, when it did not snow or sleet. And those roads, if you have never lived in Montana it is impossible to tell you how muddy those roads were. But on we went, each curve in the road bringing new scenes more beautiful than the one we left behind. Immense castles rose to meet our view while sometimes it seemed as if the mountains were great towers reaching into the clouds.

It began to grow dark and still we plodded on. But the heavy load and roads were too much for the sturdy little ponies. When we reached the Burrows' ranch there was not a pound of strength left in them. Night was coming on and certainly it was not safe to go any further.

Museum at Karst Camp. —Sara Anderson

Fortunately, we were able to persuade Mrs. Burrows (just to the south of Hell Roaring Creek) to keep us overnight—the dearest little log cabin you ever saw, only three rooms and all that mob, four others besides ourselves and the family, but Mrs. B. managed beautifully, just like all those brave women in the mountains do, and soon we were all seated at the supper table relating experiences while the little brown sorrel and the bay were being cared for at the stables.

Many times since I have remembered that home in the mountains. Mrs. Burrows, a woman of refinement and education, had seen the best of eastern life and yet was willing to leave it all for her log cabin among the mountains.

The next morning we started early—still raining but everyone seemed happy and ready to go so 'til past noon we traveled through the canyon, rising higher and higher and the river still below us racing swifter and swifter and the ever changing mountains above and around us. In the afternoon we arrived at the Michener place.

Sage Brush Point was a dangerous spot on the narrow road. The old road remains visible east of the present highway at milepost fifty-nine. When the county constructed the original road, the river ran 150 feet below at the bottom of a steep cliff. Clay, dangerously slippery when wet, covered the road. Occasionally a wagon with horses would roll down to the river. The occupants always managed to jump out and escaped serious injury. Loads, however were often swept downstream while the driver of the team worked frantically to free the horses. When cowboys drove cattle into the basin, the herders often lost a few head at Sage Brush Point, especially if it had recently rained.

Karst built up a tourist business that included guests who came for a week or two of riding and fishing. Most of his business depended on travelers visiting Yellowstone Park via the new road the county had constructed south from Taylor Fork to West Yellowstone in 1911.

Pete Karst built more than the usual spartan amenities for his guests. He had a bar and restaurant along with a museum that displayed a fine collection of minerals, Indian artifacts, a stuffed two-headed calf, and a jackalope (a cross between a jack rabbit and an antelope). He furnished his cabins with bearskin rugs and made the furniture from local lodgepole pine. He constructed the fireplaces of native stone, and hung stuffed animal heads over the hearth of each cabin.

In 1922 Karst installed a hydroelectric plant on Moose Creek and tacked a power line along the trees over to his camp, which enjoyed

Log cutting operation on Moose Creek. —U.S.D.A. Forest Service

electricity twenty-five years before the rest of the canyon had power. He installed a swimming pool and heated it with a used boiler from an old gold mine. This boiler only raised the temperature of the water a few degrees, but it provided a good selling point to dudes. Pete later held dog races on the flat land across the river, and one year the governor came down to award the dog race prizes. When Karst's stepdaughter Isabelle married Ed Durnam, a golf pro, Karst and Durnam put in a golf course across the river. Pete personally installed and maintained all the improvements his guests enjoyed.

Karst eventually had over twenty-five guest cabins clustered close to the Gallatin River. Dudes returned year after year to spend their summer listening to Pete spin tall tales. Pete prided himself on always remembering a face, treating everyone like his best friend. The first Mrs. Karst was a wonderful cook, and many Bozeman residents would drive from the valley on the weekend to have brunch at Karst Camp; only in those days they called the meal dinner. Jennie Karst died in 1931. In 1938 Pete married Nell Severence whom the Northern Pacific Railroad had sent to help Karst promote his tourist facilities.

Over the years visitors knew Pete's place by several names: Karst Camp, Karst Cold Spring Resort, Karst Rustic Camp, and Karst Kamp. Karst advertised a "fisherman's paradise, restful beds, long distance phones, mountain climbing, horses, wholesome food, dancing, dining, drinking, bonfires, curios, packtrips, Yellowstone National Park trips,

A busy week-end at Karst Camp. —Sara Anderson

The bar at Karst Camp in the 1940s. —Sara Anderson

game hunting trips for elk, deer, moose, sheep and bear, and University sorority and fraternity rush parties." Pete appealed to all people.

During prohibition locals knew Karst Camp as the place where they could openly get liquor. They had only to go into the bar and ask for a bottle of pop to get an illegal beverage. Pete remembers that "it was pretty good bootleg. No one ever went blind."

While Pete operated his own stills across the river among some rock outcrops, several other canyon residents brewed spirits for him. Margaret Michener Kelly lived on the family homestead at West Fork in 1920. She remembers the time a neighbor who had a still threw his mash over the fence. The Kelly's flock of turkeys and a milk cow reeled around for a day. Maggie mentioned to the neighbor what had happened to her animals, and he never again disposed of the mash on her side of the fence.

Pete dispensed liquor to keep the dudes and tourists happy until one day a federal agent from Idaho stopped by and asked for a drink. Pete thought he looked safe, but the agent produced a badge when Pete served him. The court sentenced Pete to spend one hundred days in the Helena jail. Pete put on such a convincing show for the court, proclaiming his remorse and explaining how his dude business would collapse without him, that the judge let him serve his term during the winter months.

Besides getting in trouble for selling liquor, Karst paid fines for illegally killing wild game. Pete once described to Dr. McGill how he had sent out a recent immigrant known as "The Big Dane" to get an elk to serve to his customers. "The Big Dane" went out and did as instructed, hauling the dead elk home in an open bob sled. The Dane didn't know how to dress out an elk, but Big Lew Bart came along and offered to do the job in return for the ivories. All of this activity came to the attention of the authorities, who brought the Dane, Big Lew and Pete before the court. Lacking the money to pay a fine, the Dane went to jail. It cost Big Lew and Pete $250 each to escape jail after the court found them guilty of failing to report the illegal kill and of aiding a criminal.

In all fairness to Pete, most people in the canyon shot elk out of season for their table. Canyon residents usually laid a green branch across the road to signify that the game warden was around. Residents also tied the horse of the visiting game warden in a spot where others could easily see it.

Karst had many colorful employees, including Walter and Luella Latta, who conducted most of the pack trips out of Karst Resort. The Lattas owned the horses and ran that end of the operation for Pete. Luella Latta had come west as a dude and stayed to marry Walter.

A 1940 brochure, put out by the Northern Pacific Railroad, described Karst Cold Spring Resort as:

> directly on the turbulent Gallatin River, fed by mountain streams, rushing through a region of snow-capped mountain peaks and billions of giant pine, fir and spruce trees. In addition to the attractive main buildings which include general store, dining room, coffee shop, lounge, and western dance hall, there are more than thirty guest cabins, most of which are modern and accommodating from one to ten persons. The cabins are located on the banks of the Gallatin River assuring cool days and nights. Many of the cabins have fireplaces and all are electrically lighted. Pete Karst has operated the Ranch for forty years and was formerly a gold prospector, guide, stage driver and is a real pioneer of the West.

The tourist business kept Pete busy, but it failed to provide him with an adequate income, so Pete went in to mining. He grubstaked many local prospectors and spent much of his own time looking for precious metals. He had most of his luck with asbestos, which he first discovered in 1916.

In an interview given about 1939, Karst tells of his mining activities:

I went out prospecting—located these asbestos claims then. We were all looking for valuable rock in those days—gold, silver, and tungsten especially—that was one of our main interests. There were lots of fellows working with me: Cliff Umdahl, Emmet Crail, Andrew Levinski, Hugh Beatty and we would get out into the mountains and put old Pat Shane to work on claims.

My mining operations were all in . . . between West Fork and Sage Brush Point on both sides of the river. All those fellows working with me were the first settlers in this community. Dick Booker and Johnny Hinckley were a couple more beside those I mentioned.

When I first located this asbestos I was at that time thinking it was worth about $100,000. Then I began to find out what it would take to handle all that stuff. I finally incorporated with some fellows from Cascade on a twenty year deal. Found out pretty soon that that also tied my hands pretty securely. So after twenty years were up I promoted it again. Starting the development of insulation. I had a young fellow whose tuition I paid down here at the college so he'd get the knowledge of how to test this stuff out for heat and whatever else had to be done. He made the analyses so I could put it on the market.

I'm not always sure of the date, the year, on these things. Years don't amount to too much when you've toled as many of them as I have.

Asbestos mine to the west of Karst Camp.
—U.S.D.A. Forest Service

Dudes riding across the bridge to the asbestos mine. —Sara Anderson

> I sold the mines not long ago—to O. E. Shepherd, used to be at the college—and C. C. Lester. And by God, not long after I sold it, they discovered the big lead that I had been looking for—that I was pretty sure was there. I'd been going in the wrong direction looking for it—they went the other way and there it was.

Shepherd and Lester opened a mine and built a road to it. They took the asbestos out on a tramway. They shipped the asbestos to Bozeman, where it is a toxic-waste problem today.

From the first rickety wagons tied together with baling wire to large taxis in the 1950s, Pete continued to provide transportation up and down the canyon. Through the years he reverted to sleds in the winter time when the county could not keep the roads open. Julia Bennett, caretaker for a time at the Nine Quarter Circle Ranch, remembers a trip that she took to Bozeman one winter in the 1930s: "I left Taylor before daylight in order to catch the stage going down. They make a trip once a week in a bobsled with a team of horses. I tied the horse behind the sled, crawled under some buffalo robes and slept all the way to Gateway."

Chuck Alderson drove the stage for many years. He used to shop for people in town and deliver their supplies on his weekly trips through the canyon. On holidays he brought boxes of food from friends and relatives in town for the prospectors who were sometimes unaware it was a holiday. During the winter, right up into the 1950s, Karst Stage delivered the mail by sled—snow often closed the road above Karst Camp to automobiles. In summer the automobiles of Karst Stage went daily from Bozeman to West Yellowstone and carried freight, mail, and six passengers.

Karst began construction of the first ski tow in Montana in the winter of 1935, and he readied the lift for use the next year. The 1,000-foot-long-wire cable towrope came from a mine, and the motor to run the lift came from a Nash automobile. Eager skiers ruined many a pair of mittens when they grasped the rough wire cable for the short ride up the steep hill.

Bozeman people flocked to the canyon on weekends to try the new sport. Young people would rent a cabin at Karst Camp for the winter at a cost of $25, and they would spend their weekends skiing and partying. Chuck Anceney, then a student at Montana State College in Bozeman, helped cajole Pete into constructing the ski run. Anceney remembers that the preferred ski boots were Lund boots from Norway and cost $12 a pair. Skis were eight feet long, made by Thor Groswald, and cost about $60 dollars a pair.

Skiers held the first organized downhill race in 1936 on Moose Creek. The next year the skiers moved the races to the new Karst Ski Run. In 1938 Chuck Anceney took first place in ski jumping in his division, first

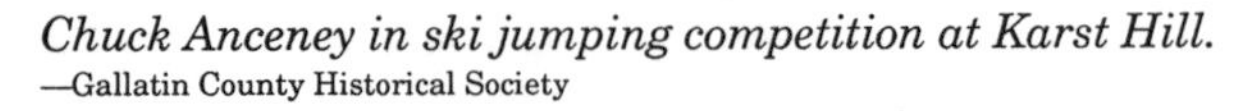

Chuck Anceney in ski jumping competition at Karst Hill.
—Gallatin County Historical Society

place in the slalom, and second place in the downhill. "I don't remember how we timed it [the downhill], but I do recall that you could fall three times and not be disqualified. One time I flew off into the trees and thought it was all over for me but I didn't get hurt and finished the race," recalls Anceney.

With the help of local enthusiasts Karst built a ski jump, which proved to be a big draw for tourists. Karst and his helpers erected a ninety-foot jump for competition and a smaller jump for practice. The local jumpers kept the run in good shape and covered the jump and the landing area with snow. George Lemon recalls filling galvanized washtubs with snow, dragging the tubs over to the jump, and packing it down.

Few Bozeman people would come up the canyon for ski races, but they showed up in droves for ski jumping. Karst held the first jumping competition in 1937. Even though the thermometer registered minus thirty degrees, 1,500 people came out to watch the event. Karst attracted the best jumpers in the country. Several Olympic jumpers showed their expertise at the ski hill, including the three Engin brothers; Kasper Oiman, who broke his leg at Karst; Adolph Peterson; and Garden Wren. These experts, mostly immigrants from Scandinavian countries, traveled around and entered any competition they could find, generously sharing their knowledge of jumping with the local boys who were new to the sport.

Chuck Anceney jumped in the BB division in 1938 and 1939. His longest jump was 140 feet. John Ring from Spokane, a jumper with a broken ankle, judged the 1938 events. Bill Penttila, a certified first-aid instructor, borrowed a toboggan and collected some blankets and splints and formed the first ski patrol in Montana. The hill died a natural death with the coming of the Second World War.

Canyon and Bozeman residents who remember Pete, love to recount stories of his activities. Buck and Helen Knight, who started Buck's T - 4 in the canyon, relate that in the thirties Pete had a large hole in the narrow road in front of his place. During the night Pete would haul water from the river and fill up the hole. During the day he would pull cars out of the hole, at $3 per car.

Pete planned to retire to a homestead on Moose Creek, but he was too much in demand by tourists to move to the solitude of the mountains. Instead he stayed on at the camp, and his step-daughter and son-in-law, Isabelle and Ed Durnam, homesteaded the Moose Creek property. Pete finally sold the camp in the early 1950s, after a half-century of catering to dudes. He spent winters in Bozeman and summers with Isabelle and Ed Durnam at their homestead. He continued to entertain friends with tales of his days in the Gallatin Canyon, a beautiful spot where a man could live free.

Thomas Michener.

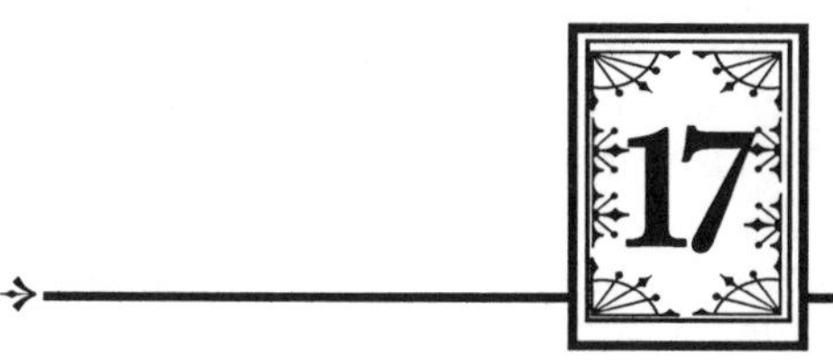

The Hercules Dredging Company

Thomas Michener was the second white child born at Alder Gulch. His father, Lewis, had left his Quaker home in Ohio and traveled west in search of gold. His mother, Emmaline Githens, had crossed the prairie with her parents and four sisters in a large wagon train. The Githens left Missouri to escape the Civil War, which killed their only son. When their plantation burned to the ground, the former slave owners headed west for a new and better life. Lewis Michener and Emmaline Githens married in Alder Gulch. They raised their son Tom on prospecting, and he searched for the earth's treasures until the day he died.

Soon after his first venture into the Gallatin Canyon in 1889 with Happy Jack Griffin, Tom became a serious student of geology. He dug enough prospecting holes to follow old river channels. He theorized that the gold he found around the West Fork came from a large deposit somewhere near Black Butte. The farther away from the source gold travels, the smaller it becomes. Gold seekers, Tom included, try to travel toward the source because the scads, or flakes, of gold, get larger, easier to find, and more plentiful closer to the mother lode. Prospectors later found gold nuggets at Black Butte.

Michener successfully followed the old riverbed of the Gallatin, a layer of gravel as wide as 120 feet when unrestricted by cliffs. He dug prospecting holes to confirm his theory that the gravel in this old riverbed contained promising amounts of gold, especially at bedrock. With this information Michener made plans to get the necessary backing to mine what he felt were rich placer deposits.

In 1911 Michener, Gladstone Stevens, and Strauss Miller incorporated under the name MS Bar. They planned to operate Michener Camp as a cattle ranch and a dude ranch, and more important to them, began to acquire land for mining. They called the mining branch of the corporation the West Gallatin Mining and Milling Company. The group owned the Michener homestead and mining claim and bought Foley Water's homestead on Porcupine Creek. Michener planned a placer-mining business as the forerunner of a much larger mining company.

Strauss Miller and Gladstone Stevens came from Portland, Oregon, to live in the canyon. Miller suffered from tuberculosis and had originally come to the Gallatin to benefit from the fresh air. He lived for a while with the Crail family and then moved to Michener Camp. His wife Della and three children joined him. Gladstone and Beulah Stevens came to the canyon after the Millers and also moved to Michener camp. The three families lived side by side in their own cabins and ate all of their meals in the main house at Michener Camp. The Micheners thought of the Stevenses and Millers as part of their family. The women took turns planning the meals and setting the table, while a hired cook prepared the meals. One winter seventeen people stayed in camp.

The men ran the ranch and dude operations and set up a method for hydraulic sluicing on the partnership's mining claims. In August 1912

Beulah and Gladstone Stevens in car with unknown man.

the *Manhattan Record* noted that: "Placer mining which has not been carried on in Gallatin County on any scale for nearly a generation is to begin again on a commercial scale in a few days in the upper West Gallatin canyon, where the West Gallatin Mining and Milling Company has about completed preparations for operation of a plant which will sluice to begin with about 1,000 cubic yards per day."

Reports vary as to how much gold the partners extracted. The scads they found ran from 960-990 fine, almost pure gold. To recover "fines" the men used mercury, but they recovered too little fine gold to pay the cost of the mercury. The lack of fine gold led them to believe that they were close to the source of the gold. This spurred the partners to begin putting together land along the upper stretches of the Gallatin River. The three men hoped that sizable tracts of riverfront land would induce investors to finance a large gold-dredging operation.

Michener, Stevens, and Miller spent the next few years putting together much of the land that covered the old riverbed from the West Fork, south for twenty-two miles, to Black Butte. In some cases they leased land from the Northern Pacific Railroad, in return, promising to pay the railroad a royalty for any minerals recovered. The minerals on United States Forest Service land were free to anyone who filed and worked a claim of which the three men filed several along the river. On private land Michener estimated how much money it would take to buy out the homesteaders. His guesses ran from $25 to $50 an acre for land along the Gallatin River.

In a prospectus for investors Michener ends his discussion of the old river bed by saying:

> If the history of gold repeats itself, the channel or streak should be narrow, and the heavier gold should all be left behind. As you follow up the channel from the lower end, the gravel, boulders, and iron rock become larger. That being so, the gold should also become more plentiful and larger; and if it increases accordingly, it will be rich enough, on bed rock, to drift mine, as Alder Gulch, Last Chance and Confederate Gulches were mined. Results of these gulches and the mining in the sixties, the world knows as history. But in none of these, was course gold found as far from the source or head of the gulch as is the course gold of the Gallatin.

Michener worked long and hard promoting gold mining in the Gallatin Canyon. With his thorough knowledge of the geology of the canyon and his careful exploration, Michener convinced several mining engineers and businessmen to form The Hercules Dredging Company to mine the Gallatin riverbed. Many of the investors came from the Spokane area, and many were mining engineers. Officers of the company included a Mr. Rheiling, who had developed the first

Incorporated Under the Laws of the State of Washington

Capital $1,000,000

All Fully Paid Common Stock Non-Assessable. (Non-Personal Liability). No Promotion Stock.

Par Value of Shares $1.00 Each

OFFICERS:

GEORGE HARDING - - President
Member of American Institute of Mining Engineers,
Manager Paulsen Realty Co., Spokane, Wash.,
Consumers Company, Coeur d'Alene, Idaho,
and Kootenai Power Company, Coeur d'Alene, Idaho.

FRANK D. ALLEN - - - Secretary
Of Allen & Allen, Mining Attorneys, Spokane, Wash.

W. A. DAVIDSON - - - Treasurer
President Gallatin Trust and Savings Bank,
Bozeman, Montana.

J. C. HAAS - Mining & Consulting Engineer
Member Amer. Institute of Mining Engineers.
Member Canadian Mining Institute.

GENERAL OFFICES: PAULSEN BUILDING
SPOKANE, WASHINGTON

PROPERTIES
Located in Gallatin County, Montana.

DEPOSITORIES OF FUNDS.
Exchange National Bank, Spokane, Washington.
Traders National Bank, Spokane, Washington.
Gallatin Trust & Savings Bank, Bozeman, Montana.

PHILIP HARDING, Fiscal Agent
301-2-3 Paulsen Building,
Spokane, Washington

Bell Phone Main 7285

A page from the prospectus for the Hercules Dredging Company.

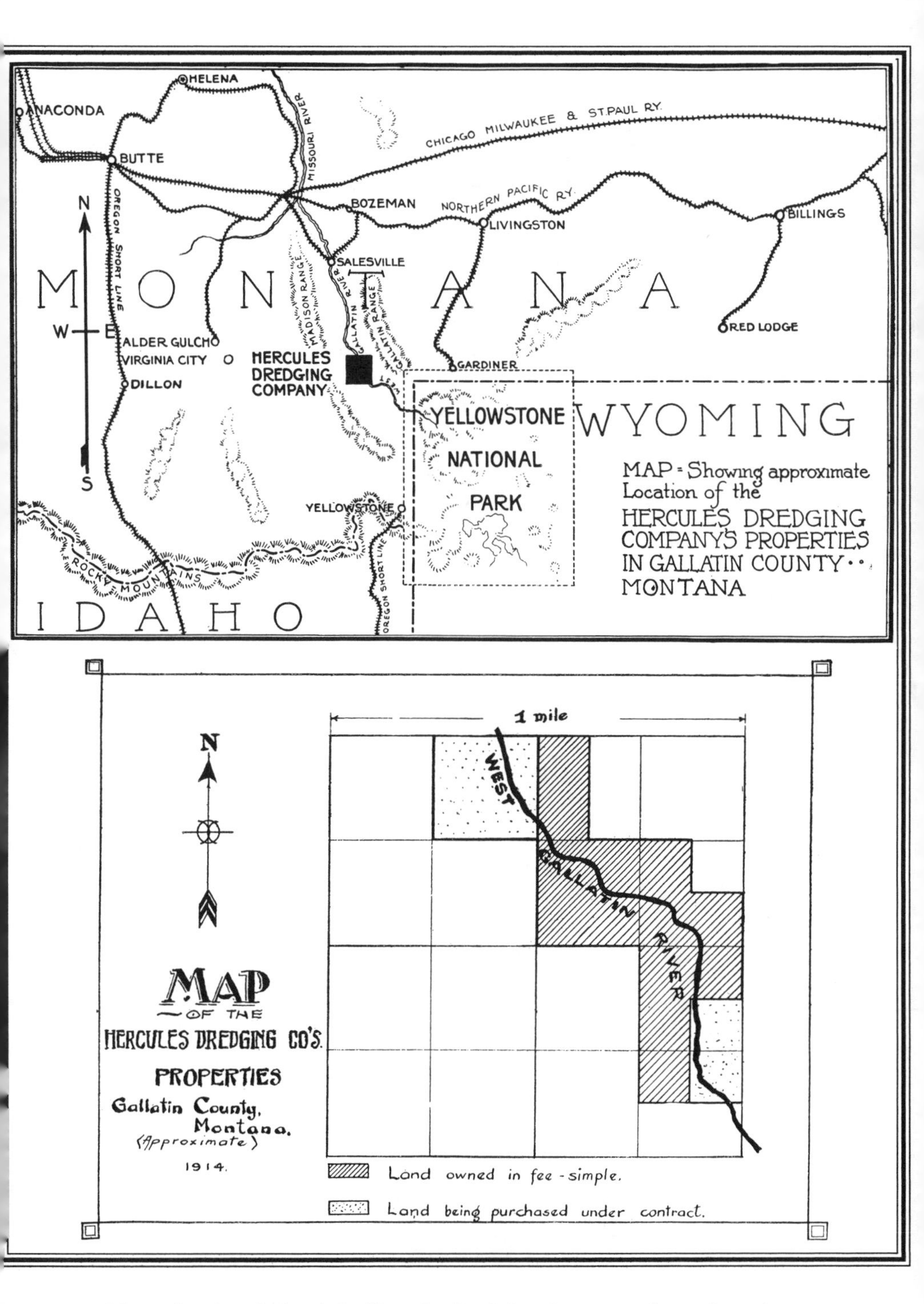

Maps showing the land the Hercules Dredging Company planned to mine.

dredge for working placer ground. Mr. Rheiling was to manage the company and to furnish the money to do a "thorough prospecting of the property and then to build and place in operation several dredges." George Harding, a mining engineer and Spokane businessman served as president, and Mr. W. A. Davidson, president of the Gallatin Trust and Savings Bank in Bozeman became the treasurer. The prospectus speaks of Thomas Michener, saying that "more than to any other man, the Hercules Dredging Company is indebted for the wisdom, enduring fortitude and integrity shown by him in the discovery, prospecting and final acquisition of its property. Mr. Michener is practically a 'born' miner and has lived his entire lifetime in the atmosphere of placer mining."

The company had authorized capital of $1,000,000, with shares selling for $1.00 each. The group planned to lease or buy up the land

Prospecting for gold with a diamond drill near Snowflake Springs.

along the Gallatin River that encompassed the rich gravel deposits. It expected to construct several large dredges, the first use of these machines in the Gallatin. The company anticipated mining from West Fork to Black Butte, concentrating on the Taylor Fork area. Coal taken from Taylor Fork would fuel machinery that would not run on water power.

In a pamphlet stating the aims of the Hercules Dredging Company, the founders of the company said:

> The Company's engineer, Mr Haas, has satisfactorily tested some of the same deposits where bedrock was reached by him in the immediate district. In doing so, he recovered values of $1.00 per bedrock yard and he has concluded that values will run considerably higher on the Hercules property. [At this time gold was pegged at sixteen dollars an ounce.] His conclusions are formed from the fact that: 1st, the Hercules property is nearer the source of the gold being further up stream. (It has been, it is believed, conclusively ascertained that the source of the gold is near the headwaters of the West Gallatin River. It is an established fact that all leads become richer and the gold courser as the source is approached). 2nd, values in the gravel above bedrock are greater on the Hercules property than at the point where this test was made. . . . The company will begin the construction of a dredge of sufficient size to economically handle its placer gravels and it is believed that the dredge will be completed early in the spring of 1915.

Unfortunately, the men started their organization as events in Europe drew the United States into World War I. The country, using more and more of its raw materials to help the Allies, experienced shortages of metals, manufactured goods, and investment capital. Hercules Dredging Company, unable to get the material to construct the large dredges and the money to finance them, put the project on hold until the end of the war.

Early in 1917 an incident occurred in the canyon which broke Tom Michener's heart. To maintain their mining claims, prospectors had to perform $100 worth of work on their claims each year and file notice of the work at the county courthouse by 31 December. If claimants failed to perform this work, other people could legally come in and "jump the claim."

On 31 December 1916 Strauss Miller and Gladstone Stevens, along with Sam Krattcer of Salesville, rode up into the mountains opposite Michener Camp to the cabin of Andrew Levinski. Miller and Stevens knew that Levinski had failed to perform the required work on his many copper claims, and they offered to buy them. Levinski demanded $10,000 for the mining rights. Miller and Stevens thought they could not pay out that much cash and countered with offers of stocks and bonds and some cash. Levinski then stated that he needed to consult his partner, Charles Anceney, Jr., before deciding. Miller and Stevens posted notice that they intended to jump the claims, but they made it clear to Levinski that their offer to buy the claims still stood. The three men left the mountain, expecting to hear within a few weeks from Levinski that he had done the work and registered with the proper authorities, or that he was willing to sell.

Levinski met with his lawyer, Walter S. Hartman, to discuss his encounter with Stevens and Miller. According to the *Bozeman Weekly Courier* of 7 February 1917, Hartman later testified that Levinski told him about the visit. Hartman told Levinski that if Miller and Stevens returned, he should refrain from shooting the two men. He advised Levinski to get to work on his claims but he recalled telling Levinski that "if they fired upon him and it became necessary to shoot to protect his life he would then have that right." After his meeting with Attorney Hartman, Levinski warned Krattcer not to show up again with Miller and Stevens, because he did not want to have to shoot him.

Andrew Levinski (standing between his two dogs at right of photo) in hunting camp. Four hunters unknown. —Christine Kundert – Burlingame Special Collections, MSU

When Miller and Stevens did not hear from Levinski, they rode into the mountains on Monday 29 January, prepared to take over the mining claims. Miller and Stevens failed to return to camp by Monday evening. Early in the morning of 30 January, Beulah Stevens, alone at Michener Camp with two of the Miller children, called Tom and Mamie Michener in Bozeman to report that the two men had not returned from the mountains. She feared a snowslide had caught them.

Tom Michener had returned the previous day from Spokane, where he had visited his doctor, and he was too ill to travel up the canyon. Tom sent his sixteen-year-old son, Charley, who attended school in Bozeman, to find the two men, though by this time everyone feared the worst. Charley Michener rode to Salesville, where he picked up Jim Noble and got fresh horses. At the Benham's place near Sheep Rock, Charles Francisco joined the two men, and they continued up the canyon.

On Wednesday morning Noble remained at Michener Camp to help Beulah Stevens with the stock. Young Michener and Francisco started up the mountain. They found both Miller and Stevens shot to death. Charley Michener untied the men's horses and put the saddle blankets over the faces of the victims. Without looking for evidence, the two men started down the mountain to notify the sheriff in Bozeman.

On Thursday Charley Michener, along with the sheriff, some deputies, and the coroner and his helper, went up the mountain to recover the bodies. They found Stevens had been shot once and Miller two times, and the sheriff found a rifle containing seven cartridges close to Miller's body. One unexploded bullet and two spent shells lay under the snow on the ground next to the body. The searchers did not find a gun near Stevens, though the undertaker later testified that he discovered a pistol under Stevens' coat. Authorities also located six more spent cartridges, both inside and outside of Levinski's cabin.

Immediately after shooting Miller and Stevens on 29 January, Levinski went to Karst Camp. He phoned his attorney, Walter S. Hartman, and asked him to report the killing to the sheriff. Levinski said that he would wait at Karst Camp for the sheriff to come to take him into custody. For some reason Hartman waited to notify the sheriff until after he had read about the murders in the Wednesday newspaper. By that time the sheriff, deputies, and coroner were already on their way to the scene of the murder, Tom Michener having told them of the killing.

Levinski spent one week in jail before his release on a $35,000 bond. Signers of the bond included Bozeman businessmen Charles Anceney, Jr., of the Flying D Ranch at Spanish Creek; Nelson Story, Jr., who had interests in the Gallatin Canyon; and others. Levinski entered a plea of not guilty, and the court set the trial for 11 June.

Deep snow hampered the search for spent bullets, and an incomplete investigation left many questions unanswered. As the snow melted, deputies and private citizens alike returned to the scene, at different times, to search for evidence. They found another gun and several more cartridges, both inside and outside Levinski's cabin. The additional evidence confirmed each finder's conclusions about the identity of the guilty party.

The Gallatin County District Court has no transcript of the trial, and all information comes from the two Bozeman newspapers. At the trial Levinski insisted that both Miller and Stevens carried guns. Jurors heard testimony that the official investigation never found a gun near Stevens, a notoriously poor shot who seldom carried a gun. Mrs. Stevens testified that her husband left the camp with snowshoes on the front of his saddle and no gun. Neighbor Henry Johnson testified that he had seen Stevens ride by with a rifle.

Levinski testified that a rifle shot from Miller or Stevens broke the window in his cabin, but investigators found all the glass from the window outside the cabin. One bullet, possibly from Miller, lodged in the ceiling of the cabin, and searchers found no other bullets inside. Levinski stated that he recognized Miller and Stevens as men associated with Michener because of the dog with them. Mrs. Stevens and Clarence Arrat, choreman at the Michener place, said that the dogs stayed in camp all day. The sheriff noted no dog tracks during the initial investigation.

During the trial, old canyon animosities surfaced. In 1914 Thomas Michener had held a mining claim to the Levinski homestead. The government disallowed this claim when they found no minerals on the land. During this dispute Charles Anceney, Jr., who for many years grubstaked Levinski while he searched for copper in the canyon, hired an attorney to secure Levinski's interest. Levinski then got title to the land under an agricultural patent.

A cast of prospectors and homesteaders hurled charges at Charles Anceney, Jr., Nelson Story, Jr., and Thomas Michener of trying to buy evidence. Through the trial Nelson Story, Jr., sat next to the defendant and frequently conferred with his attorney, Walter Hartman. Anceney paid Levinski's legal fees, and in return he received title to the Levinski homestead when the prospector disappeared two years later.

After five days of trial and a short deliberation, the jury found Levinski not guilty by reason of self defense. Within two years Levinski had vanished from the canyon, and speculation on his fate continues to this day. In 1971 Doctor Merrill Burlingame of Montana State University interviewed Rhesis Fransham, the first U.S. forest ranger in the Gallatin Canyon. Fransham, who knew Levinski well, told Burlingame that he had seen him get on a train headed east one day in 1919. Fransham repeated this story many times but never succeeded in stopping the rumors that followed Levinski's disappearance.

Gladstone Stevens and Strauss Miller lost their lives for an idea that had seduced men since the first prospectors came into the canyon. Prospectors and the men who grubstaked them believed that a vein of copper made up the mountain between Levinski Creek and Porcupine Creek. Thomas Michener pursued the quest for gold to make enough money to mine the ridge for copper. O. P. Chisholm, who had worked hard for a road up the Gallatin Canyon at the end of the nineteenth century, spent a small fortune searching for copper in the area. Rumors abounded that Nelson Story, Jr., planned to back some men in a search for copper along the fault line that runs from Levinski Creek to Porcupine Creek. These men knew that if the ridge were rich in copper, they could mine it by driving in a tunnel at the level of the valley floor. The mountain refused to release its secrets and all of these men failed in their dreams.

In 1919 Thomas Michener decided to leave the Gallatin and go to New Mexico to prospect for oil with some of his partners from the Hercules Dredging Company. Michener died in New Mexico in 1921.

Bugling bull elk with harem. —Kevin Sanders

The Gallatin Elk Herd

The influx of miners and settlers into the Rocky Mountains put pressure on the game in the region. As farmers and ranchers expanded their use of the Gallatin Valley, they began to travel into the canyon to hunt deer, elk, and mountain sheep. Newspaper reports from the 1870s recount hunting trips by valley residents who negotiated the challenging trail into the basin where they shot their winter supply of meat.

Hunters practicing the kill and eat theory put so much pressure on big game that in 1872 the territorial government set a closed season on buffalo, moose, elk, deer, sheep, goats and hare from 1 February to 15 August. As the number of small fur-bearing animals decreased, the legislature in 1876 regulated trapping. The next year the government outlawed entirely the trapping of beaver on public lands.

By 1893 concern for large game had grown so intense that the Montana legislature declared a year-round closed season on elk and moose. Legislators decided putting limits on buffalo was futile, because hunters had exterminated the herds outside of Yellowstone National Park.

Big game must have made a significant comeback. In 1895 the state Fish and Game Department, Montana's first, set limits for 1 September to 1 January of the next year. These limits included eight deer, eight sheep, eight goats, two moose and two elk per hunter. Many pot seekers came into the Gallatin canyon to fill their quota of game, which would more than keep a large family in meat over the winter. The bag limits on moose if they were hunted in the canyon would have entirely wiped out the population. Today we commonly see these large animals in the willows along the river, especially in wintertime, but at the turn of the century sighting a moose in the canyon was unusual.

As many Montana residents continued to harvest big game without thought for preservation, at the turn of the century national attention focused on the elk in Yellowstone Park. Wealthy tourists and sportsmen who had traveled and hunted in the West pressured the federal government to preserve the Yellowstone elk herds, estimated between 40,000 and 45,000 head. With the first snows, many of these animals migrate from the park to the upper Yellowstone Valley and the upper Gallatin Canyon. The elk winter in these areas and return to the park in the spring.

In 1908, responding to pressure from both national and local hunters, the United States Forest Service stopped issuing grazing permits for several areas, including lands in the Buffalo Horn-Tepee Creek drainage south to Yellowstone National Park and, on the west side of the Gallatin River, from the Taylor-Sage Divide and Shed Horn Mountain south to the park. The Forest Service wanted "to protect the elk which come out of the park to graze in winter. If horses and cattle are permitted to graze there all summer, there will be nothing left for the elk." The Forest Service continued to close areas crucial to the elk. The Northern Pacific Railroad, which owned every other section in the area, voluntarily stopped issuing grazing leases on its lands.

In 1909 Gallatin Valley sportsmen asked the United States Forest Service to set up a no-hunting buffer zone along the northern border of Yellowstone National Park. The Forest Service referred the problem to the state legislature, which established the zone in 1911. The legislature intended this district to allow the elk to disperse before they came in contact with hunters, thus stopping the pot seekers from slaughtering game as it left the park. When the elk leave the park depends on the amount of snow that falls. Early and heavy snows send the elk from the high park lands to the lower areas of the canyon in search of grass. In low snow years the elk stay in Yellowstone, and hunting is worse than in heavy snow years.

This buffer zone did not stop all illegal hunting. Roy Walton, perhaps the oldest of the old timers who used the canyon, tells of a wild hunt in 1916. In an article in the *Billings Gazette* in 1988 Walton recounted the events at "Sunnybrook Camp," located near the park boundary and named after a brand of whiskey. Thirty to forty hunters sat in camp waiting for the park elk to step over the line and present themselves to the waiting sportsmen, but the elk stayed in Yellowstone.

In desperation a group of hunters donned miners lamps to light their way through the dark forest. They entered the park, intent on driving a herd of elk toward their waiting companions. The game warden then showed up and had to be given enough whiskey to ensure that he passed out so that the illegal roundup could proceed.

The drivers, with lights on their heads, found a small herd of elk and shepherded them toward the boundary, unaware that more elk were joining the drive. Walton estimated that by the time the elk reached the hunters the herd numbered three thousand head. Walton continues the story:

> We had the other hunters fanned out on ridges up above camp to hold the elk in when we got them that far. . . . Then all hell broke loose. I don't know whatever caused the stampede. Maybe the whooping and yelling, maybe they got mad or got worried. But the elk just decided to take off.
>
> We got into more than we figured on. We picked them up in other canyons as we went through and we didn't know how many we had until the stampede.
>
> All the hunters on the ridges had their guns with them but they didn't do any shooting when the elk came over the top of them. They were scared so bad they didn't shoot.
>
> One guy, Tom Walker, was sleeping on top of a sagebrush and when he woke up, he said there were elk everywhere around him.
>
> When they broke back on us riders, they brushed my legs on each side. I can remember I rode right into them with a saddle gun and killed six head, just stuck my gun behind the shoulder and pulled the trigger.

Walton has a picture showing countless elk strung up on lodgepole logs with the largest heads positioned in front of the men.

Sensational journalism documenting the demise of the western elk kept national attention focused on the numbers of animals in and around Yellowstone Park. The continued publicity brought greater pressure on the Forest Service to protect and increase the elk herds. During the winter of 1916-17 the National Park Service, the Biological Survey, and the Forest Service conducted a census of the Gallatin elk herd to get concrete numbers. The survey determined that 1,926 elk lived in the upper and lower basins.

Even though the legislature established non-hunting zones, sportsmen continued to pressure the state to preserve additional grasslands for elk by limiting the amount of acreage devoted to grazing. By 1919 the state designated all the land on the east side of the Gallatin River above Porcupine Creek as winter elk refuge and passed legislation prohibiting cattle or sheep from grazing there. The Northern Pacific Railroad retained the land it owned in the Porcupine drainage, but it traded its grazing rights for the right to lease grasslands on Forest Service property in other areas.

During the early years of the twentieth century the Forest Service continued to move toward a policy of game management. The forestry planned to preserve the elk herds by harvesting them as a renewable resource, the policy Gifford Pinchot spelled out when the government established the Forest Service.

In 1933 Aldo Leopold outlined the measures that an agency should follow to fulfill this plan. He recommended hunting restsriction, predator control, game lands preservation, artificial replenishment (feeding hay to grazing animals), and environmental controls, including grazing restrictions. In his 1970 study of elk in the Gallatin Canyon, Allan Lovaas states that "management of the Gallatin elk herd followed this sequence almost exactly. Hunting restrictions, predator control, establishment of a National Forest and a game preserve, artificial replenishment in the form of hay feeding, and finally attempts to rehabilitate the environment by reducing grazing pressures from the elk all had their place for better or worse in the history of wildlife management in the Gallatin Canyon."

Hunters and conservationists have disagreed since the beginning of the century on how to manage the Gallatin elk herds, yet the two groups agree that preserving the elk population is crucial for the area. The groups argued over the carrying capacity of the range, and the Forest Service got the job of trying to establish the number of elk that could winter successfully in the area without supplemental feeding. Eventually, they established the number at about two thousand head.

Eric White on cross-country skis. —U.S.D.A. Forest Service

Eric White, who worked as a forest ranger in the Gallatin Canyon in the 1920s, left us with an oral history. White joined the Forest Service in 1921 and moved to the canyon in 1924 with his bride, Grace. During the summer White enforced grazing permits and marked timber for sale, and in the winter he conducted game counts. Until the 1920s Forest Service rangers, who lived in the field, served the area as game wardens.

Eric White immigrated to Montana from Pittsburgh, Pennsylvania. He served in the Army in World War I and returned to Montana to weather the extreme winter of 1919-20. Perhaps the cattle losses during that winter convinced White that he should leave ranching and look for another job. "There was an ad in the paper for a civil service forest ranger examination in Billings. I decided I would take it. . . . I'd never seen a national forest in my life. I got all kinds of requests, to go on as a ranger. I didn't know which one to take, so I picked the one closest to Yellowstone Park," White remembers. The Forest Service assigned White to the Absaroka National Forest, where he reported for work in May of 1921.

White talks about the experiences of Fred Ainger, the second forest ranger in the Gallatin Canyon:

> When Fred Ainger came in, the Forest Service had just been organized. The stockmen and the timbermen didn't accept being regulated by the Forest Service. A ranger was just an upstart. You would go into a place with everybody talking. As soon as you went in everybody clammed up. The first rangers carried guns and he had the hard job to get these people, these cattlemen, to take out permits and the timbermen to pay for the timber and it wasn't easy. When I came in that era had finished. I didn't run into a feeling against me as a ranger. On cattle, they didn't like rules I gave them, but they obeyed them. It wasn't anything like Fred had. When I left the Forest Service, we still rode horses. . . . When I came into the Forest Service you could carry the regulations in your hip pocket. . . . When I came in you were pretty much your own boss . . . and the ranger did all the work.
>
> We moved in January 1924 to the Cinnamon ranger station on the Gallatin National Forest. Its headquarters are in Bozeman. And my district then was the upper Gallatin; the Gallatin River had two ranger districts, lower and upper. Later on when cars became more proficient they were used and travel was so much quicker. They combined the districts, so later I was given the whole Gallatin River area. From Gallatin Gateway to the park.

White's duties as a ranger included issuing grazing permits, detecting and extinguishing forest fires, marking timber for sale, making game counts, and regulating hunting. And he had the authority to arrest and bring to trial those who hunted elk for their ivory tusks. He recalled:

Cinnamon Ranger Station in the 1930s. —U.S.D.A. Forest Service

There used to be a fella—he used to make a trip once a year back in that elk country at the head of the Gallatin River and he did it only for teeth. . . . We knew it but he was slick as the dickens. Nobody could ever catch him in there. . . . One time we got word that he came through there and went into that area . . . where the Elkhorn Dude Ranch is . . . back of that, up along the east park boundary. Back in that country it was in a game preserve. And so the park ranger . . . was quite interested in doing something and I was too, but the game warden . . . wasn't much interested. And old Bud Story, . . . used to be Lieutenant Governor, . . . had a cabin up there. We all went riding one day to try to catch the poacher . . . went up Monument Creek. All the way along, Bud was telling jokes, you could hear him for four blocks when he laughs. And we rode to the top of the divide and Frank and Bud says, "Oh, I don't think he came over." So, we rode back down.

This park ranger came to me afterward and he says, "You know, I think we ought to go back in that country and go in another way. I don't think we really tried yesterday, that was just kind of a circus." So we did and we got our horses and we went up Bacon Rind Creek. I was just above and we were riding along slowly, and who came out but this man with his pack outfit. He was out on the forest, the forest boundary was below us, and he had his pack horse. He was in a game preserve and carried a rifle. . . . So, we stopped him and told him we'd have to arrest him. . . . "You're carrying a gun in a game preserve. Besides that we want to search your outfit."

He unpacked his horses and we searched everything and didn't find a darn thing. But while we were doing that, he suddenly acted as if he was sick and went to where there was a big spruce tree with lots of roots sticking out. He went over there and squatted down and acted like he was sick, had his pants down . . . he said he didn't feel good. We impounded his whole outfit.

And next day, the park ranger said to me, "You know, I still wonder what he did up there, let's go up and look around." So, we went back up to where he was and looked around and in under a root where he had squatted we found his billfold, and in the billfold there was about three or four pair of elk teeth. I guess he thought we [were] going to search his body. . . . We knew we didn't have a case because of elk teeth. We'd have to prove they were killed then and not some other time. So, we also rode back up and scoured that country. We found his camp and we hunted all over to see if we could find a carcass. We gave that country an awful going over but we never did find a carcass. . . . We made an agreement with him that if he'd plead guilty to being in the game preserve with a rifle we'd try him on that basis. . . . He was convicted and willingly paid the fine, but he never got his billfold back. . . . It was very, very difficult to get a game violator convicted in this country. VERY. . . . You couldn't get a jury that would convict a man. They always favored the hunter. You had to have a lot of evidence.

Jack somebody who used to live over in Tom Miner Basin was a notorious tooth hunter. Sold them to the people in the Elk's Lodge. A good pair of teeth was worth a fortune. They'd go up to Yellowstone Park and sell them in the summertime. Jack came out over on the Gallatin for hunting and everybody knew of Jack and they watched him all the time and every time they'd run into him out on the hunting area they'd stop him and search him. He got searched so darn many times he said he was through. . . . He had a camp up on Buffalo Horn Creek. He was kind of a neighborly sort of fella and some of our men stopped there and he offered them a cup of coffee. They found out after that's where he put his elk teeth to color them, boiled them in coffee. And they were drinking that stuff . . . a lot of those old tooth hunters are gone.

Another thing, the elk came down around the station in winter. They came right up to the windows and licked the panes. One broke the window with his horns. Moose came in too in the wintertime. We put hay out for them to eat. Never did eat any; used it to sleep on.

White fed hay to the elk at the time the Forest Service practiced game management as outlined by Pinchot and Leopold.

White also studied the higher range, where domestic sheep grazed above the elk winter range, to see if it reduced elk forage. White states that no sheep grazed on winter elk range.

White and the other rangers traveled on skis or snowshoes in winter. After White became proficient on skis he preferred to travel on them, even though he had some frightening experiences:

> Lots of times I'd go out alone which isn't a good thing. I went out one time by myself and I went up past the Elkhorn Ranch on Sage Creek. I went up that creek almost to its head and there was a high divide there between Sage Creek and Monument Creek. . . . I planned to ski down Monument and then down river to Cinnamon. Monument was full of elk. . . . They dig through the snow to get down to feed and they leave great big round holes.
>
> We had to have our skis made special because they didn't have skis in those days to sell. . . . they were of ash and brittle and I walked across one of those holes and putting my weight down, my ski broke off in front of my foot and behind it and I just had a little piece on my foot. It was getting dark. I couldn't fix the ski and the snow was deep. So I decided I'd cut a branch off of the tree and see if I could fasten that on my foot like a web. It didn't work.
>
> I followed the elk trails and there was a creek there and those darn elk would go back and forth across the creek and I had to wade and was just sopping wet. . . . I borrowed a pair of skis at Story's cabin, it was a rather dangerous thing to do because my clothes froze solid. When I got to the ranger station it was about all I could do to get in. I guess that's what they call hypothermia.
>
> Nobody skied then. They weren't for sale in Bozeman. It wasn't popular like it is now. The ski is much better than a web or snowshoe. They are man-killers. There's a lot of country you can't use skis in, so you have to use webs. But the upper Gallatin was just a beautiful country to ski in. Beautiful.
>
> We learned to ski the hard way. Grace had a pair of skis for herself. We both went out, I guess it was on a Sunday or something and we got our skis on and I was ahead and I started to slide and I lost my balance and I fell down and the snow was deep. I wondered if Grace saw me and I looked around and there she was, down, too, looking around to see if I saw her.
>
> When the Montana CCCs [Civilian Conservation Corps] arrived we had them put on that elk study. They were good men and they did well on that. We built a few cabins up there with them. You could start out and stay all week over the whole basin up there. [The cabins] were stocked with food and bedding, had gas lanterns. They had little cellars dug underneath where you could put potatoes, things like that to keep. We brought horses in there in the summertime and we had this great big can with a lid on it and we had it full of rolled oats and we found that we could put eggs packed in oats and they wouldn't freeze. They kept good and you'd have fresh eggs too when you come along.
>
> The big problem was when you came into these cabins after you'd been skiing. We had stoves and everything but it took a long time to

Elk study committee. —U.S.D.A. Forest Service

get a cabin warmed and you just about froze. We had a rule that you had to leave a lot of kindling when you left and have it all ready there for when you next came in. I used to get these little cans of pork and beans. . . . I'd mix them with water and make a kind of soup, put on the heater and get it real hot. We'd drink that and that warms you up. It's a wonder some of us didn't get sick that way. Never did. Never got frostbitten. Sometimes we went ten miles. We skied from cabin to cabin . . . so we didn't have to go all the way out and back from the ranger station.

There were a few . . . ranchers near Porcupine Creek . . . just small affairs. [The elk] would bother them. . . . The trouble with the elk some winters was when heavy snows fell. Many starved to death, it was just terrible. . . . We'd make a record of all that, counting and taking photos of them.

In 1918 the *Bozeman Weekly Courier* reported that Clarence Arrat of the Gallatin Canyon had been injured while trying to drive elk away from a haystack. The article continued:

There are thousands of elk in the Gallatin district and they are hungry. Since the recent snow the ranchers have found it necessary to guard their haystacks at night to preserve them from the antlered herds. Mr. Arrat is authority for the statement that elk are dying of hunger in the lower Gallatin Basin.

> On Monday night in making the rounds he found two elk inside the corral about one haystack. He set his dogs upon them and ran to chase them out. Coming upon one suddenly as he ran around the stack, he had no time to back away before the animal reared up and jumped on him with its front hoofs.
>
> Arrat says he rolled over and tried to protect the back of his head while the elk continued to stamp upon him. He declares that the animal followed this by lying upon him in an effort to crush him, as an elk will sometimes attempt to do with a dog. He called his dogs until they left the other elk and drove the assailant away.

White did careful studies to determine how many elk the available forage would support. Encounters such as that experienced by Clarence Arrat convinced White that far too many elk grazed in the canyon. He said:

> I made an estimate of the carrying capacity of the elk winter range on the Gallatin. I estimated there shouldn't be over two thousand winter on the range and I think that was even high. . . . In the state report long afterwards they used my winter elk range capacity.
>
> Our study crew would go out and count the elk in mid-winter when they were bunched up on the range. The state didn't help. We'd make a detailed count. The sportsmen and the state wouldn't accept our count. . . .
>
> We did begin something new. We formed a committee, this was an educational movement. It was composed of representatives of businessmen and state and government officials from all around the area. Some were from the railroad, too; some were from sportsmen's organizations. In the fall we'd make a trip on horseback over all that winter elk range and look at it. Then we'd go over it again in the spring after the elk had left and show them just what the range problem was. It was a good education for all.
>
> Many of them came to the conclusion that the range was overgrazed. You see, when you don't see that range in the middle of winter you would think there was an awful lot of range up there. They didn't realize the elk couldn't get to the feed when it was covered with snow. The elk just starve to death on the small area that's available and every heavy snow winter, hundreds of them starve to death.

When White organized the committee to monitor elk, humans had eliminated the wolf, the most efficient natural predator of large game animals. The U.S. Forest Service Report on *Game Protection and Patrol of Elk Winter Range on the Absaroka and Gallatin National Forests, 1919 and 1920* urged the use of trappers in the Gallatin drainage to eradicate the predator and save the elk herds. The Forest Service put teams of wolf trappers into the head of the Gallatin River, and they must have exterminated wolves, because Eric White never mentions sighting any.

"A forest ranger lives on top of the mt. to watch for fires. Will you help him?" Sign at West Fork. —U.S.D.A. Forest Service

Fire detection and suppression kept the ranger busy in the summer. In 1910 the Forest Service began installing single-wire, grounded-circuit telephone lines in the Gallatin Canyon to enable spotters to report forest fires. Rangers strung lines from tree to tree, usually at a height of twenty feet. Sometimes elk or deer snared their antlers in the slack wires. In the early days the installers often grounded the wires with a U.S. Army World War I bayonet thrust into damp soil. They had a difficult time grounding wires on granite ridges.

Original lines went along the road or trail to make them easy to maintain, but by the 1930s the Forest Service began stringing them away from the road and out of view, if possible, so as not to mar the scenery.

A private resident could get a phone on the Forest Service line by being a "cooperator"—relaying Forest Service messages and reporting fires. The Forest Service charged $2 a year for a phone. Each party had its own ring, but every call rang in everyone's place. One long ring was central, and two longs summoned the ranger station.

In the Gallatin, installers hung the emergency phones on poles. A public phone for reporting fires hung on a tree at the West Fork, covered with glass that read, "Break only in case of fire."

White fought fires in the lower Gallatin, and said this about it:

> The Gallatin really wasn't a very heavy fire district. It was a rather wet area. It had lots of rain. I don't think that I ever had any but a little lightening fire . . . but on the lower Gallatin around the Squaw Creek Station they had fires.
>
> I went with the District Ranger to a fire at the head of Swan Creek. There were no trails. We took packs and a heavy radio. I think we took a crew with us and, oh, it was tough. Went up Swan Creek but there was so much downed timber it was just terrible. We got up there and got the fire controlled when looking around we saw another fire and went over and put it out, too. And we had radio there and we could radio back to Bozeman and they knew where we were and what the conditions were. I think we asked for some more men to come in there to mop up and be sure the fire was put out.
>
> When I was on the Gallatin I never had any lookouts or anybody patrolling for fire. We did do a little heavy patrolling on the highways on Sundays sometimes, where you had to have a shovel and axe and water bucket in your car. We'd just go along and see if people had them but that was about it. In the early days we'd go out to a fire with a crew of men. All we had to work with were axes, shovels, crosscut saws and hodacks (grub hoes) to build a trench around the fire. . . .
>
> In the late 1920s lookouts were established on Garnet, Andesite and Cinnamon Mountains.

In the early 1930s the Forest Service appointed Eric White head of the entire Gallatin Ranger District. Automobiles and an improved road made travel up and down the canyon easier, and White administered the entire district from the Squaw Creek Ranger Station. The Forest Service continued to use the Cinnamon cabin and one at Porcupine Creek for summer studies on big game and for fire prevention. At the time White moved to Squaw Creek, Jerry and Violet Noon moved into the canyon. They lived at Porcupine where Jerry worked for the Forest Service and Violet often taught at Ophir School.

In the 1930s the Forest Service strung a double telephone wire up the canyon. Although the Forest Service wanted to get out of the phone business, no phone company was willing to take over lines that stretched for thirty some miles to serve a tiny community, and the Forest Service maintained the lines for the protection of the forests until the development of the Big Sky Resort.

The Quiet Twenties

The advent of the automobile changed the way Gallatin Canyon residents lived and worked. During the summer months the automobile provided a new mobility to canyon residents and to valley people, who could spend their Sundays picnicking, fishing, and enjoying the scenery in the basin. While the first car came up the narrow dirt road in 1908, horse and wagon remained the common mode of transportation until about 1915. Residents continued to use horse and sleigh in the wintertime well into the 1920s. When snow closed the road just above Spanish Creek, canyon residents would drive a horse and sled to that point and have a friend meet them with an auto, leaving their horses at the Spanish Creek Ranch.

In 1908 Maurice Lamme, accompanied by Walter Sales, drove his new International auto from Bozeman to the West Fork. The chain-driven, high-wheeled car had hard rubber tires and a steering rod rather than a steering wheel. The men completed the news-making trip without incident. Nelson Story, Jr., had worse luck when he drove his Mitchel into the basin two years later. He struck a rock and tore off the oil pan.

Rocks and cars together continued to spell disaster. For that reason several homesteaders preferred the Dodge, because its high frame protected the underside from boulders. The first autos had gravity-fed gas tanks, which created a problem when the car went up an incline such as that at Sagebrush Point. The driver dealt with this challenge by backing up the hill, which kept the gas flowing to the engine. Still, the driver had to stop halfway up the slope to let the radiator cool off. All cars carried canvas water bags to replenish the radiator. Eventu-

Auto using the original 1898 road near Cave Creek. —U.S.D.A. Forest Service

ally cars had pressure-fed tanks, self-starters, and lights for night driving. Even with these improvements a trip in the canyon made for an adventure.

Every excursion to Bozeman involved several flat tires. The prudent traveler carried plenty of spare patches, a jack, and a pump. On one trip Billy and Pearl Lytle used all of their patches and ran the car with a flat tire for several miles. When the tire came off, they tied it back on to the rim with the baby's diaper.

In 1915 the Michener family bought an Imperial, complete with leather seats, brass trim, and carbide lamps. While impressive to look at, the car ran poorly. One day mailman Boyd Amberson's trusty Model T Ford broke down, so he borrowed the Michener's car to finish his route. The Imperial broke down near the Cave Creek Bridge and remained there, abandoned and rusting away, until a road crew pushed it over the bank. The Micheners thereafter bought Dodges and Model T Fords.

Big Lew had a Franklin that had vases inside next to the windows. At the time Lew bought the car he lived at West Fork. He learned how to start the car but not how to stop it, and the first time he drove it up the canyon he had to keep going until he ran out of gas near the park line. Someone came along, gave him some gas, and showed him how to stop the car so he could drive back home to the West Fork.

Clarence Lytle traded a cabin on Porcupine for a four-door Hudson. The car had leather seats and vases of orange carnival glass filled with artificial flowers.

Charley Sappington had a bad experience when he tried to learn to drive. After almost running over the man who turned the crank, Charley drove across the road, through a fence, and into a ditch. He walked back to the house and never drove again. When the Sappingtons

Isabelle Durnam standing on the footbridge which replaced the original pole bridge at Cave Creek, 1927. —Sara Anderson

wanted to go somewhere they had their partner Roy Harrington drive them, or they went on horseback.

Nelson (Bud) Story, Jr., and Nelson (Son) Story, III, tried many modifications on their automobiles. On some they tried to protect the underside from rock damage, and they installed heaters in others. In the 1920s they made a forerunner of the snowcat by putting a track over the wheels of an automobile to help it go through the snow, but their promising idea did not work.

One year in mid-winter Eric White and his wife Grace planned to leave the Forest Service cabin on Cinnamon and spend the rest of the season in Missoula on business. The car battery went dead, so they had to find alternate transportation. According to White,

> there was only one other way we could get out. We had a team of horses and what's called a punt, the front part of a sled, and it had a little box with a seat on it and we started out with that. It was ten below zero and snowing. We had a buffalo hide robe and a charcoal foot warmer and blankets. No track broken, the horses would only go about so long and we'd give them a rest now and then. We got as far as Squaw Creek Ranger Station. We were all day in ten below zero weather. They met us with a car there and we got into town.

Road work in the 1920s. —U.S.D.A. Forest Service

Road work in the 1920s at the Lytle grade showing both the old and the new road. —U.S.D.A. Forest Service

One May the Whites wanted to drive to Bozeman but snow still blocked the road. Eric said:

> The old ranger in my crew figured he could get out early in the morning when the snow was crusted. Grace and I and one of the other men went off with him in a Model T Ford. We started before it was light, in the morning, and drove while the crust on the snow was still hard because as soon as it was warmed by the sun you'd break through all that snow.
>
> The car had a hard time getting out. It would break through the snow and everybody'd have to get out and push and shovel and grunt in order to get it going again. Of course, a Model T, it wasn't a heavy car . . . and it was built high off the ground.
>
> After the roads opened we had to go to town for all our supplies and they [the Forest Service] had a rule that you could go only once a month. That was about the only time, even during the summer, we got out, unless you wanted to make a special trip on Sunday. . . . But, those roads were terrible and I remember coming back one time when we had to put chunks of wood in the ruts to get through because they were so deep we'd get stuck. Even with a Model T Ford it would take you all day to go down and come back. All day.

Road work in the 1920s made travel easier for the settlers and the valley people who began building summer homes in the canyon. The road work started in the early 1920s and progressed up the canyon over the next several years. In 1923 the county built the Jack Smith Bridge and moved the road to the west side of the river. It also widened and graded the road as far as Taylor Fork. In 1925 the county graveled the road from the mouth of the canyon to the park line. They extended the gravel to West Yellowstone in 1928. Those who have traveled on wet, slick, clay roads understand the importance of gravel. Gravel permitted settlers to venture out in their cars in springtime or during a rainy spell. Bud Story, who had served as lieutenant governor from 1920-24 and who

Pearl, Bert, and George Lemon at the Half-Way Inn.
—Museum of the Rockies Photo Archives

owned the Black Butte Ranch up on the park line, received the contract to do the road work.

The road crews, as they moved up the river provided business for canyon ranches. Karst Camp and the Lemon's Half-Way Inn benefited from them. The crews sometimes set up their own tent camps. Maggie Michener filled the dude cabins and ran a food service for road crews in 1923-24.

In 1919 R. B. (Bert) and Pearl Lemon arrived in a fancy Hall automobile to settle into their new home at Twin Cabins Creek. The realities of driving in the canyon soon forced them to switch to a Model T truck.

Bert Lemon grew up in Ohio, where he graduated from Ohio Weslyan University. He lived for a while in Alberta, Canada, but returned to the United States. Pearl Lemon had grown up in the mountains of Colorado, where her father practiced medicine. From

Lemon's Half-Way Inn. —George Lemon

her childhood Pearl had a knowledge of herbs and mushrooms. The couple had four children: Fern, Stanley, Wanda, and George, who was three months old when he came to the canyon. The family bought the McMullen homestead, a property containing two small, run-down cabins, each with a stone fireplace. The Lemons used one building for storing ice and the other as a tool shed, until it fell down. They named their place the Half-Way Inn and opened a store and garage to cater to the tourists driving to Yellowstone Park. Their son Stanley became a successful mechanic and built a car called "Bugs." It had a Model T motor and the body of a racing car. Stanley married Ora Michener in 1923, and they lived in Bozeman where he operated a garage.

The family cut timber and operated a sawmill, hauling lumber to the Buffalo Horn Ranch, the Nine Quarter Circle, and Bozeman. Emerson Ruegemer lived on the Safely place at Snowflake Springs and hauled lumber from the Lemons to the Elkhorn. Mr. Lemon would count the board feet and send a bill. The Millers would unload the lumber and object to being overcharged. Their complaints went on until Ruegemer moved out of the canyon, and the Millers took over the Safely place, where they found a shed full of the missing lumber.

The Lemons bought an Altman steam engine in Manhattan to use in their sawmill. It took them seven days to get it up the canyon because the engine weighed too much to be driven over the flimsy canyon bridges. The men had to ford the river every time they came to a bridge.

The family cut seventy thousand to eighty thousand board feet of lumber over the winter. They hauled lumber to Bozeman in a twelve-foot flat-bed truck. One four-by-twelve weighed 280 pounds. Cost per thousand board feet was $27 for the lumber and $25 dollars for hauling.

Charley and Maud Sappington brought cattle into the canyon in 1920. With the demise of the Cunningham and Behring operation on Taylor Fork, the number of cattle in the canyon decreased, while the number of sheep increased. The Sappingtons ran their cattle on land leased from the Northern Pacific Railroad and eventually bought several sections of land south of the West Fork. They built a house to the south of the present trailer court.

Maudie loved to entertain, and she fed her guests in grand style. She and Charley invited canyon residents for dancing on the hardwood floor of their big livingroom. The Vic Bensons of the Covered Wagon Ranch tied their organ on the running board of their touring car and drove to the Sappington's party. Maudie, entirely self taught, played fast and loud on the organ. The crowd danced until midnight, when Maudie served a full dinner. After dinner the Bensons loaded up the organ and the party ended.

Mamie Michener and the three youngest children, Helen, Ray, and Dorothy, returned in 1921 to the canyon from New Mexico, where Tom Michener had died. Margaret Michener Kelly and her husband Ray lived on the family homestead. Mamie and the younger children spent the winter of 1921-22 in the canyon and then moved to Salesville.

Ray Michener remembers that winter:

> Mother bought a cow—what a cow. She was a good milker but she was . . . I don't know the right word but crazy. She took a terrible liking to us kids and guarded us like a Doberman. I never saw or heard of anything like it. We had other cows around: Old Crip who couldn't walk good because of an overgrown hoof, Old Baldy who dearly loved to buck while people tried to ride her, and some others around there. It was something of a circus every evening. The fellows would milk the cows and then we would try to ride them. Baldy was famous for unloading people. One professional rider came by with surcingle and spurs and rode her. We should not have allowed it.

In 1917 Tom Lincoln, a crotchety old man, became the caretaker at the Nine Quarter Circle Ranch. He had previously worked for

Charley and Maude Sappington with Julie Tartar and Emmet Albrow.
—Faye Parker

Tom Lincoln at the B Bar K. —Museum of the Rockies Photo Archives

Cunningham and Behring in the Madison Valley. A misanthrope like Lincoln found spending the winter at a snowed-in ranch an ideal situation. Julia Bennett tells of an incident that she witnessed one cold winter night:

> A man came to the door and asked if he could stay all night. Tom said, "No, I am not allowed to take in anyone."
>
> And the man said, "There is a bad storm and Mr. Lincoln, I am a minister of the gospel."
>
> "Sorry Sir", said Tom, "but Jesus Christ himself could not stay here." With that he closed the door.
>
> He had a big sign outdoors that said: Have gone to Hell, will be back in a minute. I asked him what that meant, and he said that meant he was just around the corner.

Once some hunters in search of a guide called Tom, who reported to the man on the phone that Lincoln had just died. The next day a concerned friend showed up at the Nine Quarter Circle and found Tom very much alive. When asked why he said he was dead, Tom replied that he was tired of looking after hunters and couldn't think of any other way to stop the phone calls.

Tom had served in the Cold Stream Guards in England before immigrating to America, where the goldfields of Colorado lured him west. One year he had no money and decided to spend the winter in the geyser basin of Yellowstone Park. While there, Tom kept an accurate record of geyser activity that researchers found invaluable.

When Tom spent winters in the canyon, he kept a daily record of the weather. In February 1923 he made the following entries:

1st, thirty-nine below zero at post office . . .

6th, clear as a bell, colder than hell . . .

7th, damned miserable day, snowing and blowing, 22 below zero . . .

17th, first fine day in five months, 700 elk in sight.

Like the other bachelors in the canyon, Lincoln went weekly to the Eldredge Post Office at the Buffalo Horn Ranch to collect his mail.

Tom spent the last years of his life as caretaker of the B Bar K Ranch in the West Fork drainage. In the 1920s J. Fred Butler tried to buy the Nine Quarter Circle ranch, but his nephew Julius Butler beat him to it. Clarence Lytle sold his homestead at West Fork to Butler and his daughter and son-in-law, Florence and Don Kilbourne, and the Butler ranch became the B Bar K. Butler objected to sheep grazing in the drainage because of the damage they do to the land so he bought

Summer home on Forest Service land at Cascade Creek. —U.S.D.A. Forest Service

The Benson's Covered Wagon Ranch in the 1930s. —U.S.D.A. Forest Service

up seventeen sections of Northern Pacific Railroad land to keep ranchers from trailing sheep over Jack Creek. After Butler's land acquisition ranchers trucked their sheep into the canyon and continued to graze them in the West Fork drainage.

In 1914 the Forest Service issued the first special permits for summer homes at Cave Creek. Early permit holders included the Buell, Walworth, and Erwin families from Bozeman. The Forest Service issued permits for the Tamphrey Creek area in 1916. The forestry issued new leases until the late 1950s.

In 1923 the Forest Service issued permits in the Cascade area. A year's lease cost fifteen dollars. A cross section of Bozeman and Belgrade businessmen put up cabins in this popular area. Jesse Patrick built the first summer home there. Usually, carpenters used the saddle-notch form of construction for log buildings, but Patrick had grown up on the lakes of Wisconsin and wanted the butt and oakum construction found on Wisconsin lake boats. Carpenters Eugene and Emmet Crail from West Fork had the logs cut and slabbed on three sides at Karst Camp. They floated the logs down the Gallatin and used horses to pull them out at Cascade Creek. Because of its unique construction the cabin has been placed on the National Register for Historic Buildings.

Farther up the canyon Mary Sales Mills, after the death of her husband in 1927, bought three acres at Buck Creek, land once belonging to the Benhams. Tuberculosis had forced Mary to give up her homestead on Sage Creek and to go to the desert to recuperate at the turn of the century. Vic Benson of the Covered Wagon Ranch helped Mary add on to the existing one-room cabin to make a comfortable summer home. It remained without plumbing as long as Mary owned it.

Mary spent every summer at her cabin until she died in 1960. She drove up and down the road in her large black Cadillac and gave help to anyone who needed it. If her friend Josie Wilson at the Buffalo Horn Ranch needed a hand at the post office, Mary pitched in. If Mrs. Karst found herself alone when a bus load of tourists descended in search of lunch, Mary became a waitress. If the Orvises needed a hand at the rest stop, Mary sold post cards.

Mary Sales Mills' great niece Mary Owens remembers bear barbecues at the Benhams, and children piling rocks up to make a tiny swimming hole in Buck Creek.

The tall, thin, stately Mary told wonderful stories, wrote poetry, and painted scenes of the canyon in watercolors. She never missed a meeting of the Gallatin Canyon Women's Club. Her brother Reno Sales had a small cabin next to hers. After her death, Reno could not bear to go to the cabin. He sold it to Don Anderson, a newspaperman from Madison, Wisconsin, who had grown up in Bozeman.

The summer residents concerned themselves with preserving the beauty and wildness of the canyon. They tempered the demands of Bozeman residents who saw the canyon as an area they could exploit. In later years summer leaseholders sided with canyon residents against oil exploration in the canyon and against a hydroelectric dam at Spanish Creek.

In 1925 Judge Edwin Franks and Mr. V. C. Brown started a silver fox fur farm on the property that Charles Anceney, Jr., had received from Andrew Levinski. The men fenced the property, put mesh in the ground to keep animals from digging their way out, and stocked the farm with mink, muskrat, and beaver. The beaver chewed down the wooden fence posts, chased the smaller animals out, and ate all the aspens. After six or seven years the men gave up their attempts to raise fur animals in the Gallatin.

Eloise Wilson had married Bill Bennett, a government trapper, while still a student at Montana State College. The Bennetts started a mink farm on the Roan place. When the Roans sold out to Billy Lytle, Eloise moved the mink to the Clarence Lytle cabin on Porcupine. She raised the mink for breeding purposes, not for their fur. She eventually sold out and went back to college in California.

In 1927 the Milwaukee Railroad constructed the Gallatin Gateway Inn. They convinced the county commissioners to change the name of the town of Salesville to Gallatin Gateway and started sending yellow buses filled with tourists through the canyon to Yellowstone Park. The tourists needed a rest stop to break up the long ride, and Alex Shadone of the Salesville Mercantile Company received permission from the Forest Service to construct a store at a wide spot in the canyon between Buck and Cinnamon creeks. The rustic building provided the tourists with curios, candy, and postcards. Mr. and Mrs. Worth Orvis from Billings ran the store during the summer months, when the yellow buses drove up and down the canyon.

While automobiles gave canyon residents more mobility, the women of the canyon still felt isolated. They wanted to get to know each other better, and they wanted to visit without men and children around. Mrs. B. F. Freeland visited her sister in Wyoming and attended the meeting of a women's club. Her enthusiastic report spurred the women in the basin to form the Gallatin Canyon Women's Club on 21 September 1927. The women held the first meeting at the home of Eloise Bennett at the Roan homestead. The club opened its membership to all women in the Gallatin Canyon who wished to join. Dues were ten cents for every woman attending a meeting.

At the second meeting, Grace Miller, of the Elkhorn Ranch, spoke of the benefits of women's groups and discussions. At her suggestion the women listed topics that interested them, including current events, book reports, wildflower studies, and recipes. The women vetoed discussion of religion and politics. Yearly meetings began in May with a breakfast at someone's home. In mid-summer the club held a picnic for families and friends. Before winter arrived the women ended the year with a potluck dinner at one of the ranches or at the schoolhouse. The rule for the regular meetings was that the hostess for the club served dessert and beverage. When Maudie Sappington acted as hostess she provided a complete dinner of steak, chicken, fish, cooked vegetables, and ice cream with strawberries, blueberries, and raspberries served in big blue bowls.

At the final meeting of the year the women put on plays and skits, performances requiring little rehearsal. One member would read a poem or a play, and other members would act out the parts. Two popular plays were *The Burial of Blasphemous Bill* and *The Drama of Pockey Hunter*. Mary Sales Mills, Eda Benson, and Mary Owens wrote the club song to the tune of "In the Good Old Summertime."

In the good old busy time
In the good old busy time
Traveling down the gravel road with our little dimes.
You leave your work, and I'll leave mine
That's a very good sign
When we get together at the club in the good old busy time.

For the first ten years of the club the women took their accumulated dues each fall and went to lunch at the Gallatin Gateway Inn. The meal cost seventy-five cents for each member, and Mary Sales Mills always made sure that they left more than a 10 percent tip.

Grace White felt that the women should not sit idle while at the meetings, so she brought materials and had the women make quilts to send to an orphanage. The ladies made so many quilts they couldn't give them away. They finally rebelled and ceased sewing. However, they wrote letters on matters concerning the canyon, and they expected their letters to elicit a response. Over the years the women wrote to the county commissioners, the state Department of Fish and Game, the Forest Service, the governor, and the president of the United States.

To preserve the beauty of the canyon the club put trash cans along the road. Unfortunately, people used them as rubbish dumps, so the community removed the trash cans, but opened a public dump so residents could discard their old bed springs and mattresses after they finished spring cleaning.

In the early years the women had a difficult time traveling on the rough, gravel road. They often had flat tires on their way to a meeting, but most women learned how to change tires so they could be more independent. They treasured their gatherings with the other women in the canyon for the comradeship and for the knowledge they gained. Slowly the canyon was becoming less parochial.

The Depression Years

The depression made few changes in the Gallatin Canyon. The settlers had plenty of firewood and game, if little cash. They shared equipment and vehicles and worked together to bring in wood or make repairs to buildings and fences. President Hoover declared a moratorium on the work needed to maintain a mining claim or a homestead, which allowed canyon residents to leave their stakes and go to the valley to earn money. The population of the canyon remained stable—and only a few joined the community.

President Franklin Roosevelt raised the price of gold from $20 to $35 an ounce. This made placer mining an attractive activity for some young men who had just graduated from Montana State College. Feeling hopeless about getting jobs in the stagnant economy, Ray Michener and two friends moved to the Michener mining claim and began a placer operation. Ray Michener shares his memories of their experiences:

> Then, in the fall of 1933, I went gold mining, back up on the old place in the Gallatin Canyon. Raymond Lockhart, Ralph Byrne and I. . . . We set up business in the fireplace cabin and went to placer mining. We dug into a bank there near where Ray Kelly and Clarence Arrat had run a tunnel. We had about five feet of overburden, then a narrow strip of pay dirt extending down into the bedrock. We would shovel the overburden back behind us and haul the pay dirt down to the river. We didn't want to haul any rocks so we scraped them off good at the mine. What we hauled was pretty rich dirt.
>
> At first we washed our dirt in a sluice box right out in the river. It didn't work good. Billy Lytle came to our rescue. He had bought, some

time before, a little prospecting outfit. It was a real neat rig—definitely state of the art. It had a trammel screen—actually two screens—one outside the other. The inner one was course and took out the larger rocks. A self-contained pump washed the material as it was rotated in the cylindrical screens. The fines from the outer screen dropped down into the center of a centrifugal bowl that had riffles in the side. This bowl, spinning at a great rate, threw water, dirt and all, up over the edge of the bowl. Any gold was retained on the riffles. To clean up you took a plug out of the bottom of the bowl and drained the gold and a small amount of dirt into a pan. This was an "Ainley" bowl and was the best device I have ever seen for placer gold. . . .

We worked diligently until we had about three ounces of gold. You had to have $100 worth before you could send it to the mint. When we had that we shipped right away. We would continue work until the check came, and then we would go to town. We would pay up our bills, pay our respects at home, and proceed to live it up on the rest of the money. When we were broke we headed back up to the mine. . . . On one trip we had some extra money. We went over to Billings and bought a 1928 Model A Ford touring car. . . .

At that time trucks were coming over that road in increasing numbers. The road had just been worked over so they could make it from Idaho. It was still narrow and crooked. The trucks would crowd a little rig like ours clear off the road especially at night. So we rigged up a board eight foot long across the top of our windshield. We mounted running lights on this. It made us appear to be a large, wide truck. We got more respect.

We had a great time there, if you ignored the hard work. We would work until about 4 P.M., and then one of us would go fishing while the others cooked supper. This fishing was serious business. We wanted them to eat. It soon developed that I was the most dependable fisherman of the lot, so I drew that job regularly. I sometimes had to stay out pretty late but I usually brought back three fish. . . .

In the spring of 1934, when the road was opened up to West Yellowstone, a couple of men stopped by where we were starting up to wash our dirt. We had that gold machine set up in that cold spring near the big hole in the river. . . . One of these men was a foreman for Groves Construction Co.—a real big outfit. He was curious about the operation. We were busy so we let him pan some of the dirt. We were just starting up when he came up to Raymond all excited and showed him the pan with seven or eight pieces of gold. Raymond just took the pan and dumped it out on the pile. That man was taken aback if ever one was.

They stayed right there 'til we cleaned up—close to three ounces. They hung over our shoulders while we weighed it out. They asked permission to send a geologist in. We agreed so very soon here comes Pheobus. Now Pheobus wasn't a geologist, he was a promoter from Billings. He arranged for us to leave the place for one day so he and his

> helper could check it out. They paid us wages for that day. We were overjoyed. We took the day off in great spirits. Phoebus got all excited at what they found and arranged, through me, a lease with Mother. Ralph, Raymond and I were to have jobs.
>
> One of the first things Pheobus did was arrange to have water at the mine. We went up West Fork and rehabilitated Father's old diversion works. We built up the loose rock diversion dam—a wing run out into the stream. We rebuilt the upper part of the ditch that had washed out. We hired Stanley Lemon's father with a team to work on that. We didn't do a thing to the ditch below that. It just operated great.
>
> Pheobus set out right away to get a washing plant installed. He wanted to copy our little prospecting outfit in a big way. He went over to Billings and arranged for a pretty good-sized gravel screen with engine pumps and all mounted over sluice boxes. The only trouble was that the screen ran on a shaft—you couldn't put water lines inside to wash the material. We had to squirt the water in from the ends—through the sprockets in the wheels. It splattered lots of water around but did do a passable job.
>
> Groves moved in to move the dirt. . . . They moved in a bunch of horses. They weren't effective in moving that dirt, especially that bedrock. We pleaded with Pheobus and young Groves, who was managing the deal, to get in some tractors with rippers, bulldozers and such but they didn't hear us. . . .
>
> We did have a nice summer of it. Groves got quite a little gold out and I meticulously took Mother's 12 ½ percent out. She said it was the only money she had ever seen out of that claim.
>
> The only trouble was they didn't get enough gold out to pay. Old Man Groves was pretty mad about it. I told him the trouble was in their mining methods. They moved out and dropped their lease. . . . Pheobus tried to interest some other parties but he got into some stock deals in Billings and was sent to the pen for a while.

The young men took time off to look for the lost sluice box, which first entered canyon legend in the 1880s. They found an old, crude sluice box on the South Fork but failed to find gold in the nearby stream. They dug out the box made of small logs, reasoning that if there had been gold, some of it would have stuck to the logs. They took the box back to the Michener cabin and burned it in the fireplace, planning to pan the ashes for gold. But they neglected to clean the floor of the fireplace beforehand. As they panned the first ashes they heard metal scraping on the pan. Excitedly they yelled, "Gold, and lots of it." Alas, they found only some old phonograph needles from the Victrola; as the needles wore out and were replaced by new ones, the old ones went into the fireplace. That ended the lost sluice box hunt.

Before the young men stopped mining and went off to new jobs, Roy and Eric Tonn and Bailey Nixon joined in the venture. Eric Tonn married Willabelle Lytle whose family lived up the canyon.

Civilian Conservation Corps (CCC) improving the Buffalo Horn trail
—U.S.D.A. Forest Service

To help the economy and provide jobs for young men, in 1933 President Roosevelt pushed through Congress legislation establishing the Civilian Conservation Corps as part of his New Deal. The CCC provided single young men with work and training through preserving and improving the nation's natural resources. The government set up a CCC camp at Squaw Creek, making one of its major projects the construction of a new bridge there. Other contributions included

Civilian Conservation Corps (CCC) creosote treatment plant.
—U.S.D.A. Forest Service

helping on elk studies, improving backcountry trails, and constructing picnic areas and rest stops.

The corps set up a large creosote plant to treat telephone poles. The young men strung many miles of double wire as the Forest Service upgraded the telephone line in the canyon.

One New Deal scheme fortunately came to nothing. The U.S. Army Corps of Engineers proposed two possible dam sites in the Gallatin Canyon, one at West Fork and the other close to the mouth of the canyon. The Forest Service objected, and Congress failed to approve the money needed for the project. Local residents, perhaps because of the poor economy, raised little opposition. Valley residents, always ready to use the canyon for their benefit, came out in favor of a dam that would improve their farming.

In July 1932 two bandits added excitement to the placid lives of canyon residents. The men robbed a Bozeman bank early one morning, took their loot of $4,096, jumped into their sporty, green Model A Ford roadster with red wire wheels and top down, and headed into the Gallatin Canyon.

Sheriff De Vore immediately phoned the law officers in the surrounding communities, and he also called Karst Camp. A few minutes later Pete Karst called back to say that the robbers had just bought

Bridge at Squaw Creek constructed by the Civilian Conservation Corps (CCC). —U.S.D.A. Forest Service

eight gallons of gas and left. Karst reported the licence number of the car and that one of the men had a gun on his lap. A phone call to West Yellowstone assured the law men that the robbers would not escape the Gallatin Canyon.

Now the cops and robbers chase began. Paul Bohart, a young man working as a road patrolman for the Forest Service, got on his motorcycle and followed the robbers. The robbers fired several shots at Bohart, who made a quick stop at the Charley Smith ranch to borrow a rifle and five cartridges. Bohart continued the chase and caught up with the bandits just past the Benham place. The bandits stopped and fired at Bohart, who sped past as fast as he could go, weaving across the road to escape the bandit's bullets. After exchanging shots with the bandits, Bohart proceeded on to the Cinnamon Ranger Station to get more ammunition, and the robbers took to the woods close by.

Sheriffs and deputies from Bozeman and West Yellowstone converged on the upper basin. Authorities set up a command post at the 320 Ranch to organize the search for the men.

While the bandits ran into the woods, Mary Mills was visiting with the Orvises and their children at the rest stop for park buses about one mile south of where the bandits abandoned their car. Frank Latta appeared at the camp to ask the Orvises and Mrs. Mills to leave at once. He offered to drive up and down the canyon as a decoy, so they could safely reach the 320 Ranch.

If the ordinary cowboy wears a ten-gallon hat, Mr. Latta wore an eleven-gallon hat. He also sported a long curly mustache and wore a revolver strapped across his chest with an ammunition belt holding the scabbard. He had a revolver strapped to each hip, and he carried a rifle. Mr. Latta also wore a wide leather bucking belt, used by cowboys with a weak back. He carried a riata, a hand-braided horsehair rope that he said he had made one winter when he was in jail. The sight of Mr. Latta frightened Mary Mills and the Orvises almost as much as the bandits would have.

Through all the commotion Mr. and Mrs. Hincky, of Bozeman, stayed in their summer home close to the Cinnamon station. At 10 P.M., while the Hinckys were playing cards with their four house guests, the bandits entered the cabin.

After demanding a meal, the bandits gave Hincky a ten-dollar bill and instructed him to purchase food for them the next morning at the 320 Ranch. They told Hincky to put the food in the woods near the cabin. When Hincky went to the Wilson Ranch the next morning, he met several of the men from the posse. He told them the location of the bandits and begged them to stay away from his cabin for fear the bandits would kill him and the others.

Hincky returned to his cabin and put out the food. When the bandits didn't appear to retrieve the package, the Hinckys and guests made a dash for their car and sped safely off to Bozeman. Now the lawmen moved into the abandoned cabin. About 4 P.M. one bandit appeared to get the food and was instantly killed by Seth Bohart, county attorney and brother of Paul.

Sheriff De Vore, Allan Sales, Chief of Police in Bozeman, Lester Pierstorff, three other officials, and an unarmed traveling salesman looking for adventure, searched the hills north of the Hincky cabin. First in line was Sheriff De Vore followed by Sales and Pierstorff. While the search party moved toward a rocky outcrop, the robber appeared from behind some sagebrush and fired three shots at De Vore. As the sheriff fell, Sales and Pierstorff shot simultaneously and killed the robber. Sheriff De Vore's injuries were minor.

Authorities taking the dead men to Bozeman stopped along the way at each ranch to show the residents the bodies. In true western style the residents of Bozeman condemned the lawmen for killing the bandits, especially the young man who tried to get the bag of food.

The effects of the incident continued for some area residents. During the Second World War Margaret Kelly worked for the Red Cross in Portland, Oregon, taking care of wounded men. She would always ask, "Is there anyone here from Montana?" and if anyone answered "Yes," Mrs. Kelly would take them under her wing and write letters and send telegrams for them. One day a soldier came over on crutches and said, "Did you say you were from Montana?" She said, "Yes." He said, "Well, I wouldn't be very proud of it especially from that town of Bozeman." Mrs. Kelly replied that she was from Bozeman and inquired why the young man thought it was such a terrible place. He said, "Because you shot my cousin there." Startled, Mrs. Kelly asked, "My goodness, what did your cousin do?" and the soldier replied, "Well he just robbed a bank and that's no reason to shoot him."

In 1936 a remarkable woman bought the 320 Ranch. Dr. Caroline McGill had first visited the canyon twenty-five years earlier, when she joined Dr. Witherspoon and his guests in a hunting trip to Dr. Safely's place at Snowflake Springs. On another hunting trip in the early 1930s Dr. McGill stayed at both the 320 and the Elkhorn ranches.

Caroline McGill was born in Ohio in 1879 and grew up in Missouri, where she completed her life certificate for teaching at the age of seventeen. In 1908 the University of Missouri awarded her a Ph.D. in anatomy. An award allowed her to continue her studies in Europe. In 1911 Caroline McGill served as pathologist in Butte, Montana. After a year in Butte she returned east, and in 1914 she graduated from

Dr. Caroline McGill at the 320 Ranch. —Dorothy Nile

The 320 Ranch. —Dorothy Nile

John Hopkins School of Medicine, ranking first in her class. After a residency at the Mayo Clinic, the doctor returned to Butte and set up her practice.

In 1918 Dr. McGill bought a car that allowed her to explore southwestern Montana. A relaxing weekend for the doctor would find her camped along a road while she studied wildflowers, mushrooms, butterflies, and birds. When not busy studying nature, she collected antiques, which she crammed into her Butte apartment and into several storerooms.

In 1936 Dr. McGill bought the 320 Ranch from Josie Wilson. She paid Mrs. Wilson $6,000 in cash and promised to give her a home at the ranch for the rest of her life. The doctor persuaded Susie and Park Taylor, who had managed a ranch on the Madison, to move into the 320 and run the cattle and dude operation.

The cabins at the ranch had become run down, and the entire operation needed a face lift, so Dr. McGill and the Taylors refurbished the cabins. The doctor piled mattresses in her big black car and took them back to Butte to be restuffed. From her overflowing store of antiques she brought marble-topped tables and black walnut and oak furniture to the ranch. She put her collection of old clocks in her personal cabin. Silver and pewter teapots, sugar bowls, and butter dishes sat on shelves in the dining room.

The doctor spent every other weekend at the ranch, and she brought many friends to stay with her. She sent patients to the canyon to regain their health, feeling that the fresh air and good food would help them recover quickly. In 1938, Jim Flint, a mining engineer from Columbus, Ohio, electrified the ranch with a gasoline-fed generator. Flint also installed the water system at the ranch.

The doctor came to know the area through its residents. On her trips up the canyon in her big black Buick, she stopped to visit with the original settlers. She often brought them potatoes or cabbage from Logan, or sometimes apples or oranges. While she visited, she asked the homesteaders why they came to the area and how they had lived in the first years of settlement. She learned the residents interests and bought books so she could read up on the subjects. She then gave the books to the settlers. She enjoyed sharing her knowledge and love of nature.

The doctor clad her slight five-foot frame in black dress, black stockings, and a black coat with large patch pockets and a hem which ended two inches above her sturdy walking shoes. On her head she wore a tam-o-shanter adorned with a carved wooden bear. Sometimes she wore a beaver hat. In later years Dr. McGill wore a large olive drab sweater. She knitted the sweater for a friend, and when she asked him what he thought of it, he honestly said it was the ugliest looking thing he had ever seen. The doctor kept the sweater for herself and wore it for several years.

With her bright, cheery face Dr. McGill set about preserving the beauty of the canyon. She often collected roadside trash to set an example for residents, and she urged others to keep the canyon green. If she had the time, she attended meetings of the state Fish and Game

Benham Camp at Buck Creek. —U.S.D.A. Forest Service

Mabel Wood and daughter Helen at Porcupine. —U.S.D.A. Forest Service

Department and the Forest Service whenever they discussed the future of the elk in the canyon. Doctor McGill thought the future of the canyon was in tourism, and she worked hard to preserve the history and the beauty of the area.

By1938 tourists in yellow buses had stopped using the Milwaukee Railroad reststop built for their convenience. Dr. McGill bought and dismantled the building and returned the area to its original pristine condition. She used the windows from the reststop in her personal cabin.

The Benham family homestead burned in the late 1930s. Dr. McGill seized the chance to get additional pasture land for the 320 Ranch by buying up the Benham's acreage. She removed the burned buildings and returned the land to its original state. Dr. McGill gave the Benhams a few acres, so they could build themselves another cabin.

In a further bid to get pasture land Dr. McGill bought twelve sections of Holter lands on Taylor Creek in 1941, hoping to trade the Holter lands for Forest Service land next to the 320 Ranch. The Forest Service never agreed to the trade, so the Dr. leased the land on Taylor Fork to Park and Susie Taylor, who ran cattle on the land. Dr. McGill stipulated that the Taylors keep their cattle from overgrazing the area where many elk foraged in the winter.

In 1945 Dr. McGill and the Taylors bought the Jack Wood place on Porcupine Creek to raise hay. In 1950 Dr. McGill and Susie Taylor agreed to sell the Porcupine Ranch to the state Department of Fish and Game for use as elk winter range. Fish and Game assured McGill and Taylor that they would forbid any building next to the river and that fishermen would have access to the stream. During the 1940s the state acquired 6,188 acres of land—from Porcupine south to Buffalo Horn—from the Northern Pacific Railroad, paying $1.50 an acre for the land. Together with the McGill-Taylor property and a few other pieces the land became the Porcupine Game Range, a refuge used by park elk in the winter when the deep snows drive them from Yellowstone.

The doctor's keen interest in history prompted her to try to move the original Foley Water cabin to the 320 Ranch, so she could preserve the building. Unfortunately, she could not move the cabin. At one time Dr. McGill envisioned a museum built on the ranch to house the many objects of historical interest that she had collected. Instead, she founded the Museum of the Rockies in Bozeman.

In the early 1950s Patty and Jimmy Goodrich took over the operation of the 320 Ranch. In 1956 the doctor retired and spent much time at the ranch. She liked to ride alone, presenting a problem for those who felt themselves responsible for her safety. The Goodriches sent a wrangler with the doctor under the pretext of having her show the wrangler the trails around the ranch.

In her later years arthritis troubled the doctor, but she did not let it stop her from getting around, although she had to use crutches. One day she stopped to visit Dorothy Vick, who was also on crutches. As she swung out of her big car she said, "Aren't crutches wonderful? If it weren't for crutches you and I would both be in bed."

Dr. Caroline McGill died in her cabin at the 320 Ranch on 4 January 1959, at the age of seventy-nine. Besides leaving the residents a collection of papers documenting the early history of the area, the doctor left to the Gallatin Canyon Women's Club monies with which to build a meeting room. Dr. McGill always said, "If you don't have a meeting place you cannot have a community." In 1963 the women's club used the funds to build a new school that included a meeting room.

As the thirties came to a close, three families with ties to the canyon returned to the area. Isabelle and Ed Durnam began a successful venture raising lettuce for market. No longer associated with the Flying D Ranch, Chuck Anceney, his sister Rae, and her husband Deac Overturf, moved to the homestead that had come to them from Andrew Levinski. Chuck trapped and bred silver foxes. Dorothy Michener and her husband, Joe Vick, left California and moved into a cabin at Michener Camp. Joe trapped and panned for gold on the Michener family mining claim.

In 1937 Isabelle and Ed Durnam brought their new lettuce raising enterprise to the Gallatin Canyon. Isabelle, the step-daughter of Pete

(Above) Ed Durnam's lettuce operation, south of Karst Camp.
(Below) Migrant workers, Ed Durnam in the middle. —Sara Anderson

Karst, had married Ed Durnam, a golf pro, in 1923. In 1937 Ed retired from the game of golf and bought the Freeland place near Portal Creek. He and Isabelle started raising lettuce on the flat land along the west side of the river south of Karst Camp. Durnam got the idea for growing lettuce from a successful operation in West Yellowstone.

The Durnams put thirty acres into cultivation, growing the New York type of iceberg lettuce. They began planting lettuce in the sandy soil at the end of April. To irrigate the land the Durnams built a stone diversion wall, called the "wall of China," out into the Gallatin River. As many as twenty-five seasonal migrant workers hoed, weeded, picked and packed the lettuce.

In winter the family cut large blocks of ice from the Gallatin River to use in packing. In the 1940s they constructed a pond on the property where they cut ice. Covered with sawdust and kept in a shed, the ice lasted for a year. Ed and Isabelle put in a sawmill to cut wood for the shipping crates. To take advantage of the trimmings from the heads of lettuce, the family raised Hereford cattle.

The Durnams shipped their product by truck and train to markets, which stretched from Bozeman to Minneapolis. Workers packed forty-eight lettuce plants in ice in a wooden box. In a good year the Durnams shipped over four thousand crates of lettuce. The enterprise ended soon after World War II started, when labor and materials became hard to get.

In 1938 Chuck Anceney graduated from Montana State College, where he studied genetics, and moved to the Levinski homestead. The winter of1938-39 Chuck ran a trap line from Portal to Levinski Creek, checking his traps on skis every third day. It took him eight hours to climb the steep terrain and when he finished, one hour to return to the homestead. Chuck trapped marten, mink, muskrat, and weasel. At that time trappers could take beaver only if they could prove that the animals damaged their property. Anceney got a permit to take eight beaver that winter.

Anceney looked for a commercial use of his property and a way to use his training in genetics. He decided to crossbreed blue and silver foxes for the fur market. Using animals captured in Alaska, Chuck bred the foxes until he had a one-eighth cross each way. The war interrupted his plans, and his sister and her husband took over the fox farm. One night they went to Bozeman to a movie and returned to see 256 pairs of eyes staring at them from all around the property. Rae and Deac got on their horses and tried futilely to round up the foxes. Finally, Deac went into the barn and turned on the meat grinder that he used to prepare the animal's food. Two hundred fifty foxes jumped back into their cages and waited to be fed.

Deac Overturf had to get a job contributing to the war effort, so they closed down the fox operation. The Anceneys never made a profit on their venture into the fur business.

Dorothy Michener, the third person with ties to the canyon, returned in 1938. Earlier that year she had married Joe Vick in California, where they both worked in Yosemite National Park. That spring they traveled to Bozeman to visit some of Dorothy's family and decided to spend a few days at the old cabin on the Gallatin River. The few days stretched into four years.

At the beginning of the war they returned to California and worked as civilians for the Army until 1944, when they moved to the Madison Valley where Joe worked in the talc mines. Montana talc was used in components of airplanes. As soon as the war ended the couple returned to the canyon and in 1950 Dorothy Vick patented the Michener mining claim, the only patented claim in the Gallatin Canyon.

In the last years before the war, people in the canyon lived much as they always had. Some dude ranches stayed afloat by catering to people with old money. The Elkhorn was the most successful. Many young men in the canyon worked on ranches as wranglers, and a few others maintained the road for the state. Mac Blanchard, son of Frank Blanchard, had set up a sawmill where the highway maintenance building is now, and several young men worked there. The war sent these cowboys, laborers, and sawyers to places they had never dreamed of seeing. And when the soldiers and sailors returned, the canyon lurched into the modern world.

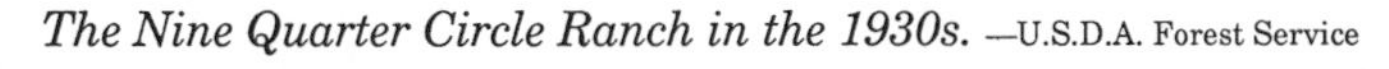

The Nine Quarter Circle Ranch in the 1930s. —U.S.D.A. Forest Service

The Stillman homestead, future home of Bucks T-4.

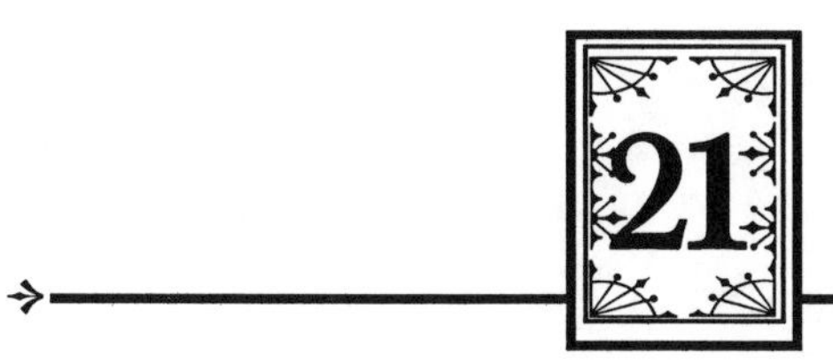

Electric Lights and Flush Toilets

World War II dominated the first half of the 1940s. Twenty-one young men left the canyon to enter the armed services. Fifteen joined the Army: Harold Adams, James Blanchard, brothers Pat and Earl Barnes, Robert Craig, Jimmy Goodrich, George Lemon, Gorden and James McMullen, Leonard and Lloyd Wortman, Jess Stovall, Russell Reckwald, Clark Taylor, and Tom Watt. Leonard Wortman was killed in Italy. Harold Adams, George Lemon, and Russell Reckwald served with the mountain ski troops. Nelson (Son) Story III, who had a ranch at Black Butte, and his son, Nelson IV, both served in the Pacific with the 163d Infantry Regiment of Montana. Nelson Story IV lost his life in New Guinea.

Billy Barnes and Vic Benson, Jr., joined the Navy. Chuck Anceney became a naval aviator. Elson Wortman and Warren Blanchard became Marines. The lone member of the Air Force was Norman Wortman.

The exodus of young men left the canyon without wranglers and sawyers, and little happened in the canyon during the first half of the 1940s. The Elkhorn Ranch continued to operate on a reduced scale, offering rest and relaxation to several diplomats and government officials.

After the war Mac Blanchard took his sawmill and moved to Gallatin Gateway. Tom Watt, Robert Craig, and George Lemon went to work for the state highway department. Several young men returned to canyon ranches to work as wranglers. Before the war Jimmy Goodrich lived on the Verwolf place, where he kept a string of pack animals, running pack trips and helping surveying crews. When

A 1940 meeting of the Gallatin Canyon Women's Club at the Ophir School.

he returned in 1945 he went to work for the Taylors taking care of the horses and cattle at the 320 Ranch. Patty Bowles worked summers at the Elkhorn, where her parents, the George Bowles, had worked for twenty-five years. In 1952-53 Patty served as teacher for three students at the Porcupine School, although she claims that she was more of a baby-sitter than a teacher. Patty's college degree was not in teaching, so she was only allowed to supervise a correspondence course for the young students. Eventually Patty married Jimmy Goodrich, and they operated the 320 Ranch for Dr. McGill, later buying the ranch under the provisions of the doctor's will.

In 1947 the residents of the Gallatin Canyon joined to get an electric line into the canyon. Montana Power declined the resident's initial overtures, so the locals threatened to go to the federal government for power under the Rural Electrification Administration. This threat caught Montana Power's attention and the company agreed to supply the residents with electricity.

Organized under the Gallatin Canyon Improvement Association, residents made and collected assessments. Herman Hintzpeter, Vic Benson and Adelaide McMullen, all of whom lived year-round in the canyon, served as officers. The assessment collectors were permanent and summer residents. Originally, in March, the association appointed seven collectors, but the job proved to be so time-consuming that they added another six collectors in July. Still, each collector had one or two difficult cases.

Summer leaseholders paid a $100 fee; permanent homeowners paid $165. Montana Power insisted that the association members take responsibility for clearing the power line right-of-way of brush and trees. The improvement association finally decided to do the work itself. The committee gave subscribers the choice of paying $40 or donating seven, eight-hour work days. Most of the summer-home leaseholders paid their assessment in cash. The permanent residents of the canyon gave their time and labor. Men cut brush and cleared trees over the winter, often in snow three-feet deep, to prepare the right-of-way for the power company.

The list of those who received electricity includes almost one hundred families. About one-quarter of these families lived permanently in the canyon while three-quarters were summer leaseholders. A few people had their own generators and objected to paying for the electric line. One or two people decided to wait for five years when they could hook up without paying the $165 assessment fee.

Grace Miller of the Elkhorn Ranch at age 82. —Barbara Hymas

When electricity finally arrived in 1949, the settlers rushed to light their homes. They installed bathrooms and kitchen sinks with running water. Many residents bought washing machines as their first appliances. Electricity lightened the workload of the men and women in the canyon.

The Gallatin Canyon Improvement Association continued to meet until 1959 when the organization dissolved. Over its life span the members established a dump, discussed the desirability of zoning, and tried to find ways to lessen the damage caused by the spruce bud worm.

The question of permitting oil drilling shows the way the people in southwestern Montana perceived the Gallatin Canyon. Bozeman businesses in the late 1940s supported Phillips Petroleum Company in its bid to drill for oil on Pika Mountain, in the Wapiti drainage of Taylor Fork. Canyon residents and summer leaseholders voted to oppose oil exploration on public lands. The Forest Service granted

The Elkhorn Ranch dining room. —Barbara Hymas

easements to the company, and it drilled for oil but found insufficient reason to press on with its drilling beyond 2,500 feet. Phillips Petroleum left the canyon and paid the Forest Service to restore the road it had cut into the land.

A U.S. Reclamation Bureau plan in the 1950s to build a dam at Spanish Creek had approval from Bozeman residents, but those in the canyon opposed it. Once again, the government decided against carrying through with its proposal.

Dude ranching revived with the prosperity the country enjoyed after the war. Howard Kelsey returned from the Army and bought the Nine Quarter Circle Ranch. He made extensive changes and additions, and built up a large cliental. Wranglers returned to the canyon, and dudes once more rode out of the Elkhorn, the 320 Ranch, and Karst Camp.

Dorothy and Joe Vick made some money by allowing dudes to pan for gold at their place at West Fork. The Vicks gave the visitors a pan,

The corrals at the Elkhorn Ranch. —Barbara Hymas

Patty and Jimmy Goodrich at the 320 Ranch. —Dorothy Nile

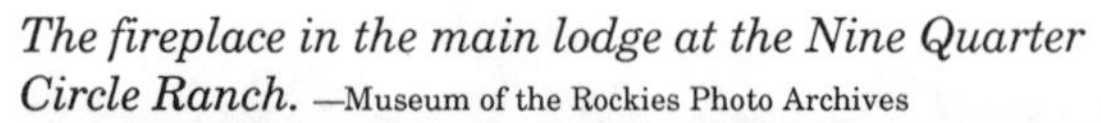

The fireplace in the main lodge at the Nine Quarter Circle Ranch. —Museum of the Rockies Photo Archives

To the right, Dorothy and Joe Vick at their gold panning operation.

Below, tourists panning gold at the Michener mining claim. The man on the right with the sketch book under his arm is the American artist John Steuart Curry, known for his depictions of the plains.

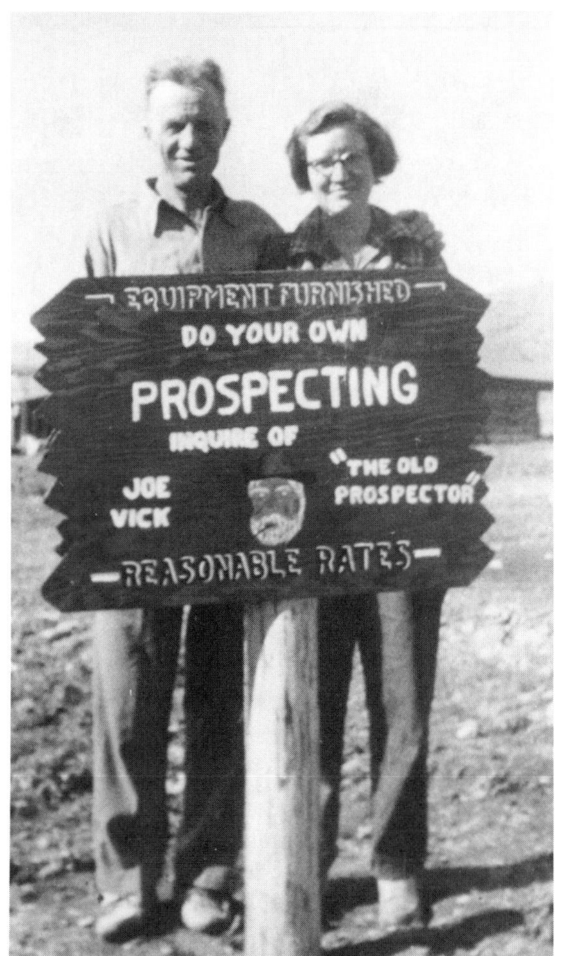

The Frank Blanchards at their home on Dudley Creek. Note the slab roof.
—Museum of the Rockies Photo Archives

a pick, and a shovel, and told them where to dig. The first dudes to try their hand at gold panning, the mayor of Cincinnati and three school teachers, came from Karst Camp. They found $8 worth of gold in one hour. The Vicks weighed the gold on scales brought to Alder Gulch in 1865 by Lewis Michener.

Three new businesses began at the end of the 1940s. Buck and Helen Knight moved from Livingston and bought the Bert Stillman homestead for $10,000. During the 1930s Buck's dad had wrangled for the B Bar K. Buck had lived with his aunt and uncle, Jack and Mabel Wood, at Porcupine and had attended the Ophir School. The Knights built several log cabins along the road and opened Buck's T - 4, a bar and restaurant. They housed road crews in their cabins and took in hunters and fishermen.

Farther up the canyon Alma and Art Vandecar opened the Almart Lodge, with a gas station, restaurant, and cabins. They catered to the tourist driving the canyon after a trip through Yellowstone Park. Max Spore built the Corral Bar, across from Twin Cabins Creek, in the late 1940s.

The end of the war brought new building materials to the canyon. Residents had previously made cabin roofs of slab wood, wood shingles, or tar paper and dirt. Now asphalt shingles came on the market, and almost every homeowner decided he or she needed a new roof. Travelers in the canyon in the late 1940s must have thought a hurricane had just gone by. Every house had a brand new roof or a pile of asphalt shingles in the yard. Nails, hinges, and other metal fixtures and tools allowed the settlers to make repairs that they had postponed for the duration of the war.

During the 1940s F. O. Butler and his daughter Florence Kilbourne died. Their heirs sold the B Bar K Ranch to Bob Turner and Earl Reiser, who used the ranch as a camp for young men. Turner, a minister, wanted to instill values and leadership skills in the young men he envisioned as the future leaders of the United States. Most of the campers came from well-to-do families, but Turner offered scholarships to a certain number of young men unable to pay the fee. The camp operated for a few years, and then Turner and Reiser sold the B Bar K to Don Corcoran, a business man from Bemidji, Minnesota. Corcoran had stayed at Buck's T - 4 while hunting, and he knew the area.

Corcoran started a major pulp-wood operation in the West Fork drainage. He brought in Chippewa Indian lumberjacks and cut timber from seventeen sections he had acquired from Turner and Reiser. He also cut timber on national forest lands. The men started cutting sections close to the B Bar K, spreading out from there. The loggers cut the timber to eight-foot lengths, then used horses to skid the logs to the roads where workers loaded them on trucks. From Gallatin

The B Bar K Ranch, now the Lone Mountain Ranch. —Vivian Schaap

Horses returning to the B Bar K Ranch. —Vivian Schaap

Gateway, Corcoran shipped the logs via the Milwaukee Railroad to Minnesota.

The Corcorans lived at the B Bar K Ranch. They housed workers in tarpaper shacks spotted all around the drainage. Each shack measured nine feet by twelve feet. The company allotted a family with four or more children two shacks for living. One year, seventy-eight school age children lived in the area. The Corcorans paid the school teachers and held classes at the ranch. The pulp operation lasted about five years, and like the Cooper Lumber operation in the early 1900s, it ceased abruptly. A saturated pulp market reduced the demand for wood and Corcoran closed down his business.

A paved road resulted from the pulp operations. Trucks carrying the logs to Bozeman needed a better road and the U.S.Bureau of Public Roads and the U.S. Forest Service obliged by working on the road for much of the 1950s. Road crews changed the course of the Gallatin River in several spots to accommodate the new and improved road, eliciting angry protests from residents. Most objected to the increased truck traffic on the narrow, winding road, and many people felt that the blasting and dirt work spoiled the beauty of the canyon. They saw the road improvements as an example of outside interests dictating the way they lived.

In 1950 Jack and Elaine Hume, who owned and operated restaurants in Chicago, moved to the West Fork area. Jack Hume had stayed at Buck's T - 4 while hunting. The Humes originally bought the 960 acre Crail property, running cattle and setting up a sawmill to supply canyon people with lumber. In 1955 they bought 900 or so acres from Don Corcoran, who retained the seventeen sections that he had logged over. The property acquired by the Humes included the B Bar K Ranch, which the Humes renovated and opened as the Lone Mountain Ranch. A partner in the Humes operation, Tom Boa, worked for Marshall Field in Chicago, and he referred many dudes and hunters to the ranch. The Humes ran the Lone Mountain Ranch until 1962, when they sold it and all of their other acreage to Sam and Florence Smeding.

Through the 1940s the Forest Service leased lands up Beaver Creek and in the Yellow Mules, southeast of the West Fork meadows, to sheep ranchers from the Madison Valley. Dude ranchers who used these areas for pack trips and riding objected to this practice, and the Forest Service finally phased out sheep leases.

Canyon residents became a tightly knit group through the Gallatin Canyon Improvement Association. By the beginning of the 1950s a consensus had formed that the area needed a church. Accustomed to providing for themselves, the residents pitched in together. One offered land at the mouth of the West Fork. Several others promised logs, and some offered time and expertise.

Son Story heard that the residents planned to build a chapel. He told them that in his will he had provided for a chapel at West Fork. Son and his wife Velma planned to dedicated the chapel to the memory of their son Nelson Story IV and the eighty-one other members of the 163d Infantry Regiment of Montana who had been killed in the Second World War.

While grateful to the Storys for their intention, the residents wanted the chapel in the present, not the future. Son Story joked that he was afraid to drive up and down the canyon for fear someone would run him off the road, so in 1954 the Storys gave the land and the money to build the chapel. In a speech about his plans, Son said, "we are going to build a memorial chapel at the mouth of West Fork Creek in Gallatin Canyon. . . . The building will be of heavy stone, log and shake construction in harmony with the mountain setting . . . and will be oriented on Lone Mountain as a reminder of the 121st Psalm. . . . The Montana soldiers who died for the defense of our freedom were of every faith. It is our wish that the worship conducted in this chapel be nonsectarian. . . ."

Fred Willson and Harry Grabow, Bozeman architects, drew the plans, and I. M. Johnson worked as contractor. Russell Rehm and his sons, who recently moved into the canyon, did the stone work. Jimmy Goodrich, from the 320 Ranch, and Dick Miller, from the Elkhorn, cut the logs and delivered them to the building site. Dorothy and Joe Vick donated the land for the long driveway into the chapel. Three thousand people attended the day-long dedication of Soldiers Chapel on 2 October 1955.

The chapel drew the members of the community together and provided an opportunity for the settlers to work side by side. Three community members serve on the thirteen person board of directors.

Many events and shared concerns helped to gather the canyon residents together into a community. The automobile and an improved road gave them greater mobility, allowing the settlers to see more of each other, to get together and socialize, and to trade ideas. The desire for electricity and a common concern about their environment brought the settlers and summer homeowners closer together. The Gallatin Canyon Improvement Association gave the residents a place to air their hopes and their concerns. The Soldiers Chapel and the new Ophir School provided the residents with places to meet. In giving money to the Gallatin Canyon Women's Club, Dr. McGill had stressed the need for the community to have a meeting room.

The number of year-round residents remained small, yet most of these families realized the need to preserve the beauty of the canyon. They debated the fate of the elk herds and fed the animals when the snows were deep. They welcomed the expansion of dude ranches as a clean and profitable use of the land. They liked the life they lived in the canyon. Residents and summer people agreed that the area should be saved for recreation. Right through the 1960s the inhabitants of the canyon lived a sheltered, quiet life.

In 1968 Chet Huntley, evening newscaster for NBC, vacationed at the 320 Ranch. Chet was a charismatic dreamer who had grown up in Montana. Though he already owned a ranch in central Montana, he confided to Jimmy and Patty Goodrich that he wanted to buy a cattle ranch in the mountains. He planned to put together a group of investors who would use the ranch for a tax write off and as a place for vacations. Huntley asked the Goodriches to look for a place for him to buy. Jimmy Goodrich suggested the Lone Mountain Ranch and additional acreage in the West Fork drainage.

How a proposed cattle ranch turned into a year round resort makes a story for another book. For better or worse, Gallatin Canyon residents, who thought their area should be preserved for recreation, were about to be presented with recreational development well beyond what they had anticipated, the Big Sky Resort.

Dedication of Soldiers Chapel, October 2, 1955.

Bibliography

Anceney, Charles L. III. *The Anceneys of Meadowbrook and the Flying D.* Bozeman: Gallatin County Historical Society, 1986.

Bennett, Julia Bembrick. Merrill G. Burlingame Special Collections. Accession 2023. The Libraries, Montana State University.

Bicentennial Oral History Project. Audio Tape Storage. Merrill G. Burlingame Special Collections, The Libraries, Montana State University.

Bryan, William L., Jr. *Montana's Indians Yesterday and Today.* Helena: Montana Magazine, 1985.

Burlingame, Merrill G. "Gallatin Canyon Development Traced" (in three parts). *Bozeman Chronicle,* 15-17 February 1970.

———. Interview with Christine Kundert. Merrill G. Burlingame Special Collections. Accession 1151. The Libraries, Montana State University.

Coats, J.H. *Communications in the National Forests of the Northern Regions.* U.S. Department of Agriculture, Pamphlet 1984.

Comin, Katherine. *Economic Beginnings of the Far West.* New York: The MacMillan Company, 1912.

Cooper, Miriam Bunker. Merrill G. Burlingame Special Collections. Accession 2183. The Libraries, Montana State University.

de Lacy, Walter Washington. *Contributions to the Historical Society of Montana,* Vols. 1 and 2. Rocky Mountain Publishing Company, 1876.

DeVoto, Bernard, ed. *The Journals of Lewis and Clark, by Meriwether Lewis and William Clark.* Boston: Houghton Mifflin Company, 1953.

Dominick, David. "The Sheepeaters." *Annals of Wyoming* 36 (October 1964).

Garcia, Andrew. *Tough Trip Through Paradise.* Edited by Bennett H. Stein. Sausalito: Comstock Edition, 1976.

Gates, Paul Wallace. *History of Public Land Development.* Washington, D.C.: Government Printing Office, 1968.

Graves, Henry, and E. W. Nelson. *Our National Elk Herds.* Washington, D.C.: U.S. Department of Agriculture, Circular 51, 1919.

Haines, Aubrey L. *The Bannock Indian Trail.* Yellowstone Library and Museum Association, 1964.

Hawkes, Lewis E. "Gene." *Reading the Forest, A History of the Natural Resources Issues in the Headwaters of the Gallatin, Madison, and Yellowstone Drainages and the Gallatin National Forest.* Unpublished Manuscript. Bozeman, U.S. Forest Service, n.d.

Howard, Joseph Kinsey. *Montana: High, Wide, and Handsome.* New Haven: Yale University Press, 1943.

Hultkrantz, Ake. "The Indians of Yellowstone National Park." *Annals of Wyoming,* 29 (October 1957).

Janetski, Joel C. *Indians of Yellowstone Park.* Salt Lake City: University of Utah Press, 1987.

Koch, Elers. *Early Days in the Forest Service.* Vol. 1, Missoula: U.S. Department of Agriculture, Forest Service, Northern Region, 1944.

Lamar, Howard R., ed. *Reader's Encyclopedia of the American West.* New York: Crowell, 1977.

Lovaas, Allan L. *People and the Gallatin Elk Herd.* Helena: Montana Fish and Game Department, 1970.

Malone, Michael P., and Richard B. Roeder. *Montana: A History of Two Centuries.* Seattle: University of Washington Press, 1976.

Marble, Charles. *Diaries of Charles "Buckskin Charley" Marble.* Merrill G. Burlingame Special Collections. Accession 430. The Libraries, Montana State University.

McGill, Caroline. Papers. Merrill G. Burlingame Special Collections. Accessions 925, 945, and 945a. The Libraries, Montana State University.

Michener, Thomas. Personal papers, privately owned.

Michener, Raymond. Personal papers, privately owned.

Napton, Lewis K. "Canyon and Valley, Preliminary Archaeological Survey in the Gallatin Area, Montana." Masters Thesis, University of Montana, 1966.

Phillips, Paul C. *The Fur Trade.* Vol. 2. Norman: University of Oklahoma Press, 1961.

Russell, Osborne. *Journal of a Trapper.* Edited by Aubrey Haines. Lincoln: University of Nebraska Press, 1955.

Snyder, Gerald S. *In the Footsteps of Lewis and Clark.* Washington, D.C.: National Geographic Society, 1970.

Thorson, Margaret Michener Kelly. Personal papers, privately owned.

Todd, Bayard. *Salesville and the Todd Family.* Bozeman: Gallatin County Historical Society, 1984.

Toole, K. Ross. *Montana: An Uncommon Land.* Norman: University of Oklahoma Press, 1959.

White, Eric. Interview by Laurie Mercier. Montana at Work, Oral History Project for the Montana Historical Society, 1983.

Wilkins, Lela Shepherd. "The Sedgwick Benham Family and Sheep Rock Ranch." Prepared for the Centennial History Conference, Bozeman, Montana, 1989.

Index